AF615751

SOME ITERATIVE SOLUTIONS IN OPTIMAL CONTROL

SOME ITERATIVE SOLUTIONS IN OPTIMAL CONTROL

John B. Plant

RESEARCH MONOGRAPH No. 44
THE M. I. T. PRESS, CAMBRIDGE, MASSACHUSETTS

Set by UNEOPRINT *and printed and bound in the United States of America by The Riverside Press*

Library of Congress catalog card number: 67-27348

To Kay

Foreword

This is the forty-fourth volume in the M.I.T. Research Monograph Series published by the M.I.T. Press. The objective of this series is to contribute to the professional literature a number of significant pieces of research, larger in scope than journal articles but normally less ambitious than finished books. We believe that such studies deserve a wider circulation than can be accomplished by informal channels, and we hope that this form of publication will make them readily accessible to research organizations, libraries, and independent workers.

Howard W. Johnson

Acknowledgments

The computations reported here were carried out either at the M.I.T. Computation Center or the Computing Center of the Royal Military College, Kingston, Ontario, and the author is grateful for the cooperation he received at both centers.

Preface

Few, if any, optimal-control systems have been built. There are several reasons for this state of affairs, the first being that the solutions to time-optimal-control problems are extremely complicated. A second reason is that the currently known solution methods are, with few exceptions, iterative in nature, and these are, for the most part, characterized by unknown rates of convergence. To make things worse, even less is known about the problem when the plant is nonlinear or noisy. There is even some question as to whether the practicing engineer would wish to synthesize an optimal controller as such. Why, then, other than as an academic exercise, should one publish a work entitled *Some Iterative Solutions in Optimal Control?*

A little reflection will convince even the most skeptical that algorithms for computing optimal controls can be extremely useful to the practicing engineer even if he has no desire to build an "optimal" system. The control engineer's *raison d'être* is the improvement of performance: How can one be sure that a significant improvement can be made on an existing design without solving the optimal-control problem? The engineer is often required to choose between several designs: this choice can be made on a more rational basis if it is known how the performance of these designs compares with the optimal performance in the light of several performance criteria. The algorithms presented in this work can be used for these purposes.

It is conceivable that methods will be developed in the future for synthesizing suboptimal controllers using known solutions to various optimal-control problems. The designer using these schemes will require algorithms for solving optimal-control problems.

Much of the research reported here was carried out while the author was a doctoral candidate at the Massachusetts Institute of Technology. The work was done under the supervision of Professor Michael Athans, and the author wishes to express his sincere gratitude for the excellent advice and encouragement that he received during his candidacy. Special thanks are extended also to Professors Roger W. Brockett and George Zames, who acted as readers of the author's thesis and who also gave guidance, and to Professor George C. Newton, who was the author's Faculty Counselor. Finally, the writer is endebted to Professor M. Blythe Broughton of the Royal Military College of Canada for his assistance in editing the manuscript.

Kingston, Ontario, Canada
October, 1966 *John B. Plant*

Contents

SOME ITERATIVE SOLUTIONS IN OPTIMAL CONTROL

1. Notation, Terminology, and Concepts

1.1 Introduction

The purpose of this chapter is (1) to establish notation, (2) to define formally some pertinent control-theory notions such as a dynamical system, controllability, an optimal-control problem, etc., (3) to state the maximum (minimum) principle of Pontryagin, and (4) to define the classes of problems with which this work is concerned.

1.2 Notation[1]

Vector notation and set-theory concepts are used throughout. Sets are denoted by upper-case letters, with or without subscript; the elements of a set are designated by lower-case letters. A set S composed of all objects x characterized by some property P is defined by a statement having the following abbreviated form:

$$S \equiv \{x : x \text{ has property } P\}. \tag{1.1}$$

The symbols of set algebra used here are defined in Appendix A. The names and properties of special sets such as compact, convex, function, etc., are also to be found in Appendix A. Vectors are denoted by boldface lower-case letters and matrices by boldface upper-case letters. The transpose of a vector $\mathbf{x}$ and that of a matrix $\mathbf{A}$ are denoted by $\mathbf{x}'$ and $\mathbf{A}'$, respectively. The Euclidean n-Space (the set of all real n-dimensional vectors) is denoted by R_n.

---

[1] The notation and terminology conforms, for the most part, to that found in Reference 3. See also Reference 23.

The scalar product of two vectors $\mathbf{x}$ and $\mathbf{y}$ in R_n, the Euclidean norm of a vector $\mathbf{x}$ in R_n, the quadratic form of a matrix $\mathbf{A}$ in $R_n \times R_n$, and the dyad of two vectors $\mathbf{x}$ and $\mathbf{y}$ in R_n are denoted by $\langle \mathbf{x}, \mathbf{y} \rangle$, $\|\mathbf{x}\|$, $\langle \mathbf{x}, \mathbf{A}\,\mathbf{x} \rangle$, and $\mathbf{x}\rangle\,\langle \mathbf{y}$, respectively. These terms are defined more precisely in Appendix B. The notation $\mathbf{x}(t)$ is taken to mean the valuation at time t of a vector-valued function of time $\mathbf{x}_{(0, t_1]}$. In other words, $\mathbf{x}_{(0, t_1]}$ is a function from $(0, t_1]$ into R_n, and $\mathbf{x}(t)$ is its value at time t in $(0, t_1]$. The time derivative of $\mathbf{x}(t)$, the gradient of a scalar-valued function of many variables $L(\mathbf{x}, t)$ with respect to $\mathbf{x}$, the partial derivative of $L(\mathbf{x}, t)$ with respect to t, the first derivative of a vector-valued function $\mathbf{f}(\mathbf{x})$ with respect to $\mathbf{x}$ (a matrix), and the second derivative of $\mathbf{f}(\mathbf{x})$ with respect to $\mathbf{x}$ are denoted by $\dot{\mathbf{x}}(t)$, $L_{\mathbf{x}}(x, t)$, $L_t(\mathbf{x}, t)$, $\mathbf{f}_{\mathbf{x}}(x)$, and $\mathbf{f}_{\mathbf{x}\,\mathbf{x}}(x)$, respectively. These terms are defined in Appendix B.

1.3 Some Pertinent Control Theory Notions[2]

Definition 1.1 (Dynamical System)

The following aggregate of mathematical objects is called a dynamical system: a set Ω of real m-vectors $\mathbf{u}$ (that is, $\Omega \subset R_m$); the Euclidean n-space of real n-vectors $\mathbf{x}$ (that is, $\mathbf{x} \in R_n$); an interval $(t_0, T]$ in R_1; a function $g(\mathbf{x}, u, t)$ from $R_n \times R_m \times R_1$ into R_l; and a vector differential equation

$$\dot{\mathbf{x}}(t) = \mathbf{f}(\mathbf{x}(t), \mathbf{u}(t), t), \tag{1.2}$$

for which, given an initial condition $\mathbf{x}(t_0) = \mathbf{x}_0$ and a piecewise continuous function $\mathbf{u}_{(t_0, T]}$ there exists a unique absolutely continuous function $\mathbf{x}_{(t_0, T]}$ which satisfies Equation 1.2 almost everywhere in $(t_0, T]$. The vector differential equation, Equation 1.2, is called the plant. A function $\mathbf{u}_{(t_0, T]}$ with $\mathbf{u}(t) \in \Omega$ is called an admissible input over the observation interval $(t_0, T]$. The solution of Equation 1.2 is called a state trajectory, and $\mathbf{x}(t)$ is called the state of the plant at time t. The function $y_{(t_0, T]}$, where

$$y(t) = g(\mathbf{x}(t), \mathbf{u}(t), t), \tag{1.3}$$

is called the output. Let $U((t_0, T]; \Omega)$ be the set of piecewise continuous functions on $(t_0, T]$ with values in Ω (that is, $\mathbf{u}(t) \in \Omega$ for t in $(t_0, T]$), the set $U((t_0, T]; \Omega)$, or simply U, is called the input space.

---

[2] These concepts are discussed more completely and in more generality in Reference 3, Chapter 4, and in Reference 41.

Consider Equation 1.2. The uniqueness of the solutions to this equation implies the existence of a function $\psi(t; \mathbf{u}_{(t_0, t]}, \mathbf{x}(t_0))$ such that

$$\mathbf{x}(t) = \psi(t; \mathbf{u}_{(t_0, t]}, \mathbf{x}(t_0)). \qquad (1.4)$$

Here ψ is called the plant transition function. The reader will find a more complete discussion in Reference 3 (pp. 159–167).

Definition 1.2 (The Optimal-Control Problem Z)

The problem has five conditions. (1) Let U be the input space of a given dynamical system in which the initial state $\mathbf{x}(t_0) = \mathbf{x}_0$; (2) let $L(\mathbf{x}(t), \mathbf{u}(t), t)$ be a continuous function from $R_n \times R_m \times (t_0, T]$ into R_1 (T not necessarily given); (3) let $F(\mathbf{x}(T))$ be a once differentiable function from R_n into R_1; (4) let

$$J(\mathbf{x}_0, t_0; \mathbf{u}_{(t_0, T]}) = F(\mathbf{x}(T)) + \int_{t_0}^{T} L(\mathbf{x}(s), \mathbf{u}(s)s)\, ds \qquad (1.5)$$

and

$$\xi(\mathbf{x}_0, t_0; \mathbf{u}_{(t_0, T]}) = \int_{t_0}^{T} h(\mathbf{x}(s), \mathbf{u}(s), s)\, ds \qquad (1.6)$$

be performance and constraint functionals, respectively, from $R_n \times R_1 \times U$ into R_1; and (5) let $S \subset R_n$ be a target set. Then the problem is to find the control $\mathbf{u}_{(t_0, T]}$ in U that "steers" the state from $\mathbf{x}_0$ at t_0 to S, such that J (Equation 1.5) is minimized and (Equation 1.6) satisfies the relation $\xi \leq \hat{\xi}$, where $\hat{\xi}$ is assumed given, is called the optimal-control Problem Z in this work.

The optimal-control problem is usually formulated without the constraint introduced by Equation 1.6 and $\hat{\xi}$ (that is, usually $\hat{\xi} = \infty$). This constraint is included here for the sake of generality, and because it is part of a class of problems treated in Chapter 3.

Several names have been given to the various special cases of Z in the literature. For example, the term "regulator problem" is used when S is the single point at the origin (that is, $\mathbf{0} \in R_n$); if $S = R_n$, the problem is said to be "free-end point"; if S is a vector-valued function $\omega(t) \subset R_n$ and the problem is to have $\mathbf{x}(t) = \omega(T)$, then the term "rendezvous" is used; if, however, it is required that $\mathbf{x}(t) = \omega(t)$ for $t \geq T$, then the problem is one of "tracking." Neustadt has coined the phrase "minimum effort" for problems where $L = L(\mathbf{u}(t), t)$ and $F = 0$. These minimum-effort problems include the "minimum-fuel" $\left(L = \sum_{i=1}^{m} u_i\right)$ and the minimum energy $\left(L = \frac{1}{2} \sum_{i=1}^{m} u_i^2\right)$ problems; in the "minimum-time" problem, $L = \text{const}$ and $F = 0$; and finally, if $L = 0$ and $S = R_n$ one uses the term "terminal cost."

It is assumed in this monograph that either $L = 0$ and $S = R_n$ or $L = L(\mathbf{u}(t), t)$, $F = 0$, $\hat{\xi} = \infty$, and $S \neq R_n$. These two types of problems are closely related and, as will be pointed out, a sequence of solutions to the first problem can result in the solution of the second.

A control $\mathbf{u}_{(t_0, T]}$ that satisfies the constraints (that is, $\mathbf{u}(t) \in \Omega$, $\mathbf{x}(t_0) = \mathbf{x}_0$, $\xi(\mathbf{x}_0, t_0; \mathbf{u}_{(t_0, T]}) \leq \hat{\xi}$, $\mathbf{x}(T) \in S$) but that does not necessarily minimize the cost J is called a feasible control.

If the optimal control exists, it and all of the parameters associated with it are denoted by a superscript asterisk, that is, $u^*_{(0, t_1^*]}$, $\mathbf{x}^*(T^*)$, J^*, etc.

Definition 1. 3 (Linear Plant)

A plant described by an equation such as Equation 1. 2 in which both of the functions **g** and **f** are linear in both **x** and **u**, is said to be a linear plant. That is, we have

$$\dot{\mathbf{x}}(t) = \mathbf{A}(t)\mathbf{x}(t) + \mathbf{B}(t)\mathbf{u}(t), \tag{1.7}$$

$$\mathbf{y}(t) = \mathbf{C}(t)\mathbf{x}(t) + \mathbf{D}(t)\mathbf{u}(t), \tag{1.8}$$

where $\mathbf{A}(t)$, $B(t)$, $\mathbf{C}(t)$, and $\mathbf{D}(t)$ are $n \times n$, $n \times m$, $r \times n$, and $r \times m$ matrices, respectively.

It is well known ([3], pp. 171–172) that there is a state-transition matrix (also called a fundamental matrix) $\boldsymbol{\phi}(t, t_0)$ associated with the plant of Equation 1. 7, such that one has

$$\mathbf{x}(t) = \boldsymbol{\psi}(t, \mathbf{u}_{(t_0, t]}, \mathbf{x}(t_0)) = \boldsymbol{\phi}(t, t_0)\ (\mathbf{x}(t_0) + \int_{t_0}^{t} \phi^{-1}(s, t_0)B(s)\mathbf{u}(s)\ ds), \tag{1.9}$$

where $\boldsymbol{\phi}(t, t_0)$ is the solution of

$$\boldsymbol{\phi}(t, t_0) = \mathbf{A}(t)\phi(t, t_0), \qquad \boldsymbol{\phi}(t_0, t_0) = \mathbf{I}, \tag{1.10}$$

where **I** is the $n \times n$ identity matrix and (see Reference 41)

$$\boldsymbol{\phi}^{-1}(t, t_0) = \phi(t_0, t). \tag{1.11}$$

This monograph is concerned primarily with linear plants.

Definition 1. 4 (Reachable State)

A state $\mathbf{x}_1 \in R_n$ is reachable from a state $\mathbf{x}_0 \in R_n$ at time t_0 and with respect to the input space U and with $\xi \leq \hat{\xi}$, if there exists a $\mathbf{u}_{(t_0, T]}$ in U that steers the state from $\mathbf{x}(t_0) = \mathbf{x}_0$ to $\mathbf{x}(T) = \mathbf{x}_1$ with T finite and with $\xi \leq \hat{\xi}$.

Definition 1. 5 (Controllability)

Let the state $\mathbf{x}_1$ in Definition 1. 4 be the origin **0** of R_n. If $\mathbf{x}_0 \in R_n$ and the state **0** is reachable from $\mathbf{x}_0$ at time t_0 in finite time T and with respect to some input space U and with $\xi \leq \hat{\xi}$, then $\mathbf{x}_0$ is said to be controllable at t_0. If every $\mathbf{x}_0 \in R_n$ is controllable at every $t_0 \in R_1^+$,[3] then the plant is said to be completely controllable.

Definition 1. 6 (Domain of Controllability)

Let $C(t_0)$ denote the set of all states $\mathbf{x} \in R_n$ from which the origin is reachable at time t_0, for a given plant, a given input space, and a given $\hat{\xi}$. The set $C(t_0)$ (or simply C in the time-invariant case) is called the domain of controllability of the plant with respect to Ω, ξ, and $\hat{\xi}$.

Definition 1. 7 (The Set of Reachable States R)

Let $R(\Omega; T, \mathbf{x}_0, \hat{\xi}) \subset R_n$ denote the set of all states that are reachable from $\mathbf{x}_0$ in time $T \geq t_0$ with $\mathbf{u}(t) \in \Omega$ for $t \in (t_0, T]$ and $\xi \leq \hat{\xi}$; namely,

$$R(\Omega; T, \mathbf{x}(t_0), \hat{\xi}) = (\mathbf{x}:\mathbf{x} \in R_n, \mathbf{x} = \psi(T; \mathbf{u}_{(t_0, T]}, \mathbf{x}(t_0)), \tag{1.12}$$

$$\mathbf{u}(t) \in \Omega \quad \text{for} \quad t \in (t_0, T], \quad \text{such that} \quad \xi(\mathbf{x}_0, t_0; \mathbf{u}_{(t_0, T]}) \leq \hat{\xi}).$$

Theorem 1. 1 ([3], Theorem 4. 2)

Let the plant be linear and constant, that is, let

$$\dot{\mathbf{x}}(t) = \mathbf{A}\mathbf{x}(t) + \mathbf{B}\mathbf{u}(t), \tag{1.13}$$

with $\Omega = R_m$ and $\xi = \infty$, and let b_β, $\beta = 1, 2, \ldots, m$ denote the βth column of $\mathbf{B}$. Set

$$\mathbf{e}_{\alpha\beta} = \mathbf{A}^\alpha \mathbf{b}_\beta, \quad \alpha = 0, 1, \ldots, n-1, \quad \beta = 1, 2, \ldots, m, \tag{1.14}$$

and let E be the set of n-vectors

$$E = \{\mathbf{e}_{\alpha\beta}:\mathbf{e}_{\alpha\beta} = \mathbf{A}^\alpha \mathbf{b}_\beta, \alpha = 0, 1, \ldots, n-1, \beta = 1, 2 \ldots, m\}. \tag{1.15}$$

Then, the plant of Equation 1. 7 is completely controllable if and only if there are n linearly independent vectors in E.

Definition 1. 8 (Controllability Matrix)

The $n \times n$ matrix $\mathbf{C}_\beta$, whose columns are the vectors $\mathbf{e}_{\alpha\beta}$, $\alpha = 0, 1, \ldots, n-1$ of Theorem 1. 1; that is,

$$\mathbf{C}_\beta = [\mathbf{b}_\beta \,\vdots\, \mathbf{A}\mathbf{b}_\beta \ldots \mathbf{A}^{n-1}\mathbf{b}_\beta], \tag{1.16}$$

[3] R_1^+ = positive half of the real line.

is called the controllability matrix of u_β. The β subscript is omitted if $m = 1$.

Definition 1.9 (Observability)

The state $\mathbf{x}_0$ at time t_0 of a given plant is said to be observable at time t_0 if, given any control $\mathbf{u}_{(t_0,T]}$, there is a time $t_1 \in (t_0, T]$, perhaps dependent on $\mathbf{u}$, such that a knowledge of $\mathbf{u}_{(t_0,t_1]}$ and the output $\mathbf{y}_{(t_0,t_1]}$ is sufficient to determine $\mathbf{x}_0$. If every state is observable at t_0, then the plant is said to be completely observable.

It is assumed throughout this work that the plant is completely controllable and completely observable and that the given initial state $\mathbf{x}_0$ in Problem Z is in the domain of controllability $C(t_0)$.

1.4 The Minimum Principle

The Maximum Principle of Pontryagin, henceforth referred to as the Minimum Principle, is the most powerful theorem[4] available for the solution of optimal-control problems of the type considered in this work. Actually, the theorem applies, formally speaking, only for $\hat{\xi} = \infty$ (that is, no effort constraint) in that such constraints do not appear in Reference 4. However, the conditions of the theorem can be used effectively to gain an insight into the solution of the problem when $\hat{\xi}$ is finite also. The proof of the theorem, first published in English in Reference 4, is also discussed in detail in Reference 3. Though the theorem is not rigorously proved in Reference 3, this book is recommended to engineers without the mathematical background required to follow Reference 4. The results of the theorem are simply stated here without proof.

Theorem 1.2

Consider Problem Z. Let $\hat{\xi} = \infty$, and let $\mathbf{f}(\mathbf{x}, \mathbf{u}, t)$, $\mathbf{f}_\mathbf{x}(\mathbf{x}, \mathbf{u}, t)$, $L(\mathbf{x}, \mathbf{u}, t)$, $L_\mathbf{x}(\mathbf{x}, \mathbf{u}, t)$, $L_t(\mathbf{x}, \mathbf{u}, t)$ be continuous on $R_n \times R_1 \times \Omega$, and $H(\mathbf{x}, \mathbf{u}, \mathbf{p}, p_0, t)$ be a continuous function from $R_n \times R_m \times R_n \times R_1$ into R_1 defined as

$$H(\mathbf{x}(t), \mathbf{u}(t), \mathbf{p}(t), p_0, t) = p_0 L(\mathbf{x}(t), \mathbf{u}(t), t) + \mathbf{p}(t), \mathbf{f}(\mathbf{x}(t), \mathbf{u}(t), t). \tag{1.17}$$

If there is an optimal control $\mathbf{u}^*_{(t_0,T^*]}$ in $U((t_0, T^*], \Omega)$, then there

[4] Dynamic programming [9], [12] is also used extensively in optimization problems, but it is not useful in the type of problems treated here.

necessarily exists a constant p_0^* and an n-vector $p^*(t)$, such that

NC 1. $p_0^* \geqslant 0;$ (1.18)

NC 2. The (n + 1)-vector $\begin{bmatrix} p_0^* \\ \text{----} \\ \mathbf{p}^*(t) \end{bmatrix} \neq 0;$ (1.19)

NC 3. $\dot{\mathbf{x}}^*(t) = H_{\mathbf{p}}(\mathbf{x}^*(t), \mathbf{u}^*(t), \mathbf{p}^*(t), p_0^*, t);$ (1.20)

NC 4. $\dot{\mathbf{p}}^*(t) = H_{\mathbf{x}}(\mathbf{x}^*(t), \mathbf{u}^*(t), \mathbf{p}^*(t), p_0^*, t);$ (1.21)

NC 5. $\mathbf{x}^*(t_0) = \mathbf{x}_0;$ (1.22)

NC 6. $\mathbf{x}^*(T^*) \in S;$ (1.23)

NC 7. Along the optimal trajectory $\mathbf{x}^*(t_0, T^*]$

$H(\mathbf{x}^*(t), \mathbf{u}^*(t), \mathbf{p}^*(t), p_0, t) \leqslant H(\mathbf{x}^*(t), \mathbf{w}(t), \mathbf{p}^*(t), p_0^*, t),$
for all $w(t) \in \Omega;$ (1.24)

NC 8. If the plant is time-invariant, then

$H(\mathbf{x}^*(t), \mathbf{u}^*(t), \mathbf{p}^*(t), p_0^*)$ is constant on $(t_0, T^*];$

NC 9. If T is not specified (that is, if T is "free"),

$H(\mathbf{x}^*(T^*), \mathbf{u}^*(T^*), \mathbf{p}^*(T^*), p_0^*, T^*) = 0;$ (1.25)

NC. 10. If the target set S is a smooth (k + 1)-fold in $R_n \times (t_0, T]$ (that is, if there are (n – k) functions $g_1(\mathbf{x}, t), g_2(\mathbf{x}, t), \ldots, g_{n-k}(\mathbf{x}, t)$, $1 \leqslant k \leqslant n - 1$, such that

$S = \{\mathbf{x}, t) : g_i(\mathbf{x}, t) = 0, i = 1, 2, \ldots, n - k\})$

then the vector $\mathbf{p}^*(T^*)$ is an outward normal to S at $\mathbf{x}^*(T^*) \in S$ (that is, $\langle \mathbf{p}^*(T^*), \mathbf{x} - \mathbf{x}^*(T^*) \rangle = 0$ for all $x \in T(\mathbf{x}^*(T^*))$, where $\mathbf{T}(\mathbf{x}^*(T^*))$ is a support hyperplane of S at $\mathbf{x}^*(T^*)$;

NC 11. If the target set S is the entire state space R_n, then $\mathbf{p}(T) = F_{\mathbf{x}}(\mathbf{x}(T))$. (Note: it is assumed that $F = 0$ if $S \neq R_n$).

The vector $\mathbf{p}(t)$ is called the <u>costate</u>, and its associated differential equation (NC4) is called the <u>adjoint system</u>.

If $p_0^* \neq 0$, one may set $p_0^* = 1$ (see [3] Chap. 5), and the problem is often called <u>regular</u>.

Definition 1.10 (Normal)

Let $H(\mathbf{x}, \mathbf{u}, \mathbf{p}, p_0, t)$ be the Hamiltonian of Problem Z. If $H(\mathbf{x}, \mathbf{w}, \mathbf{p}, p_0, t)$ considered as a function of $\mathbf{w}$ has a unique minimum

$$\mathbf{w} = \mathbf{W}(\mathbf{x}, \mathbf{p}, p_0, t) \tag{1.26}$$

for all $\mathbf{x}, \mathbf{p}, p_0$, and t such that $\mathbf{W}(\mathbf{x}(t), \mathbf{p}(t), p_0, t)$ is uniquely defined almost everywhere in (t_0, T), then Z is said to be normal, $\mathbf{W}(\mathbf{x}, \mathbf{p}, p_0, t)$ is called the control law, and the ensuing $\mathbf{w}_{(t_0, T]}$ is called an extremal control. If Z is not normal, it is said to be singular.

The problems considered in this work are assumed to be normal.

The regulator problem is perhaps the most practical subset of the optimal control problems in Z. Certainly this problem has been treated extensively in the literature. This problem also receives considerable attention here in a modified form. The target $\mathbf{0}$ is replaced by a target set S_1 where

$$S_1 = (\mathbf{x}: \langle \mathbf{x}, \mathbf{Q}\mathbf{x} \rangle \leqslant r^2), \qquad r > 0, \qquad \langle \mathbf{x}, \mathbf{Q}\mathbf{x} \rangle \geqslant 0 \forall \mathbf{x}. \tag{1.27}$$

It is argued that this problem is, in many cases, a more meaningful mathematical model of the engineering regulator problem when r is small, and it is shown that this formulation, called the "modified regulator problem," leads to some computational and mathematical advantages in the use of a particular algorithm.

One other class of problems is treated here in detail. This problem is a more general one from the point of view of restrictions on the function L and the set Ω and is called "the fixed interval, free-end-point, terminal-cost, effort-constrained, optimal-control problem." It is shown in Chapter 3 that successive solutions of this problem can yield the solution to many others.

In particular, this monograph is a study of the following five subsets of Z.

Definition 1.11 (Problem Z_1)

Let Z_1 denote the subset of Z in which (1) $\dot{\mathbf{x}}(t) = \mathbf{A}(t)\mathbf{x}(t) + \mathbf{B}(t)\mathbf{u}(t)$; (2) $L(\mathbf{x}(t), \mathbf{u}(t), t) = 1$; (3) $\Omega = \Omega_a = (\mathbf{u}: |u_i| \leqslant 1, i = 1, 2, \ldots, m)$; (4) $F = 0$; (5) $S = (\mathbf{x}: \langle \mathbf{x}, \mathbf{Q}\mathbf{x} \rangle - r^2 \leqslant 0)$, $r > 0$, $\mathbf{Q}$ positive definite; and (6) $\hat{\xi} = \infty$. Problem Z_1 is called the "time-optimal (Ω_a)" or the "minimum time" Ω_a modified regulator problem.[5]

[5] An early treatment of this problem can be found in Reference 10.

Definition 1.12 (Problem Z_2)

Let Z_2 be the subset of Z defined as for Z_1 (Definition 1.11), except that $\Omega = \Omega_b = (\mathbf{u}: \langle \mathbf{u}, \mathbf{Ru} \rangle \leq 1)$, where $\mathbf{R}$ is a positive definite $m \times m$ matrix with $m > 1$. Problem Z_2 is called the "time-optimal (Ω_b)" modified regulator problem.

Problem Z_1 is normal if and only if all of the controllability matrices $\mathbf{C}_\beta$, $\beta = 1, 2, \ldots, m$ are nonsingular, whereas Problem Z_2 is normal if any of the controllability matrices $\mathbf{C}_\beta$ is nonsingular (Appendix C).

Definition 1.13 (Problem Z_3)

Let Z_3 be the subset of Z defined as in Z_1 (Definition 1.11), except $T > t_0$ is fixed and $L = \sum_{i=1}^{m} |u_i|$. This problem is called the "fixed-time fuel-optimal" modified regulator problem.

Problem Z_3 is normal if all of the controllability matrices $\mathbf{C}_\beta$, $\beta = 1, 2, \ldots, m$ and the matrix $\mathbf{A}$ ($\mathbf{A}$ assumed constant) are nonsingular (Appendix C).

Definition 1.14 (Problem Z_4)

Let Z_4 denote the subset of Z defined as in Z_3, except that $L = \sum_{i=1}^{m} \frac{1}{\rho} |u_i|^\rho, \rho > 1$. Note that Problem Z_4 becomes Problem Z_3 if $\rho = 1$, and it becomes Z_1 if $\rho = 0$ and T is free. However, these problems are treated separately, since they differ in many respects.

Problem Z_4 is normal.

Definition 1.15 (Problem Z_5)

Let Z_5 denote the subset of Z for which (1) $\dot{\mathbf{x}}(t) = \mathbf{A}(t)\mathbf{x}(t) + \mathbf{B}(t)\mathbf{u}(t)$; (2) $\Omega(t) \subset R_m$ is convex and compact; (3) $S = \mathbf{z} \in R_n$ (a constant n-vector); (4) T is fixed; (5) $L(\mathbf{x}(t), \mathbf{u}(t), t) = 0$; (6) $F = \frac{1}{2} \langle \mathbf{x}, \mathbf{Qx} \rangle$, where $\mathbf{Q}$ is positive definite; and (7) $h(\mathbf{x}(t), \mathbf{u}(t), t) = h(\|\mathbf{u}(t)\|)$, where h is monotone increasing and convex down, and $\|\mathbf{u}\|$ means any norm of $\mathbf{u}$.

Let T^* denote the minimum time for Problem Z_1. Since T^* is the minimum time in which the state can be steered from $\mathbf{x}_0$ to the target, Problem Z_3 and Z_4 will have no solution unless the fixed time $T \geq T^*$. Further, if $T = T^*$ and the time-optimal control is unique, Problems Z_1, Z_3, and Z_4 will have precisely the same solution.

Consider Problem Z_5, and let T^* and $\bar{\mathbf{u}}_{(t_0, T^*]}$ denote the least terminal time and the time-optimal control, respectively, for the problem of steering the state from $\mathbf{x}_0$ to $\mathbf{z}$ with $\mathbf{u}(t)$ in Ω. Assuming that $T > T^*$, let $\hat{\mathbf{u}}_{(t_0, T]}$ denote a control which steers the state from $\mathbf{x}_0$ to $\mathbf{z}$ and minimizes the constraint functional ξ. Then, if $T \leq T^*$, it is assumed that $\hat{\xi}$ (Definition 1.2) satisfies $\hat{\xi} \leq \xi(\bar{\mathbf{u}}_{(t_0, T^*]})$, and if $T > T^*$, $\hat{\xi} \leq \xi(\hat{\mathbf{u}}_{(t_0, T]})$.

2. Background

2.1 Introduction

This chapter is a survey of some of the currently known methods of solving some special cases of the optimal-control problem. Two known noniterative solutions are given in Section 2.2, and their limitations are discussed. The remainder of the chapter is spent on iterative methods, and these are divided into four categories. The first method, due to Ho [19], is a successive approximation of the time-optimal regulator problem by a fixed-time terminal-cost problem. The second method, proposed by L. W. Neustadt, is based on the convexity (for linear plants) of the set of states that can be reached from a given state with fixed cost. Though Neustadt has extended his method to include minimum-effort problems [31], it is presented here only in terms of the time-optimal regulator problem. The third method has been proposed by N. N. Krasovskii and R. V. Gamkrelidze. The last method uses the first terms of Taylor-series expansions to improve the guess. The gradient method and Newton's method fall into this category. We discuss the work of H. K. Knudsen, R. W. Bass, and B. Paiewonsky in the application of these methods to the time-optimal problem. The results reported here for the modified regulator problem also fall into this category. In fact, we show that some of the difficulties of these approaches are removed when the problem is modified.

2.2 Noniterative Methods

The noniterative solution is the most desirable because if such a solution exists, the optimum control can be expressed as a function of the time and the current state of the plant.

In Problem Z_1, the time-optimal control is "bang-bang," and in Problem Z_3, the fuel-optimal control is piecewise constant at levels $+1$,

-1, and 0. The piecewise constancy of these controls leads naturally to the subdivision of the state space into regions of constant control. The hypersurfaces which subdivide the state space in this manner are called "switch surfaces." For time-invariant systems of low order ($n \leqslant 3$, $m \leqslant 2$), a feedback controller can be constructed utilizing these surfaces [3], [21], [26], [35]. These controllers use nonlinear time-invariant function generators to approximate the intersection of the switch surfaces with the coordinate planes of the state space. A typical example that illustrates the type of hardware required may be found in Reference 3 (Figure 7–26). When $n = 2$, the switch surfaces can be found experimentally, even if $\mathbf{f}(\mathbf{x}, u)$ (assumed time-invariant) is unknown [3], [26].

Another noniterative solution is obtained for a special class of problems in References 4 and 5. The result is independent of n but is restricted to plants of the form

$$\dot{\mathbf{x}} = \mathbf{f}_1(\mathbf{x}, t) + \mathbf{u}, \tag{2.1}$$

$$\|\mathbf{u}\| \leqslant 1, \tag{2.2}$$

for which one has

$$\langle \mathbf{f}_1(\mathbf{x}, t), \mathbf{x} \rangle = g(\langle \mathbf{x}, \mathbf{x} \rangle, t) \tag{2.3}$$

for all $\mathbf{x}$ and t.

When n is large, the switch-surface approach becomes unusable because equations describing the surfaces cannot easily be found as explicit functions of the state. Also, there are no published noniterative solutions to Problem Z_4. Consequently, one must turn to iterative techniques for the solution of these problems. The fact that optimal-control problems require iterative solutions is disappointing, particularly if the solution is required in real time (that is, while the plant is in operation). Nevertheless, an iterative solution is still important for several reasons. In most applications, the suboptimal-control scheme is the practical answer to the inability to express solutions to optimal-control problems in closed form (that is, yielding the optimal control as an explicit function of the state). These schemes are engineering approximations to, and/or compromises between, the solutions of a multitude of optimal-control problems (that is, a multitude of performance criteria). However, given a performance criterion, there is only one way to find out how close a given suboptimal design is to being optimal with respect to that criterion, and that is to solve the optimal-control problem. This "yardstick" capability of iterative solutions is useful to the designer in a rather indirect way. Iterative solutions to optimal-control problems can be useful in a more direct manner in that they can be used to construct closed-loop suboptimal-control schemes. An obvious, but perhaps not too sophisticated, approach might be to represent the region of interest of the state space by a grid of discrete points. An iterative procedure could be used to find the optimal control for each point.

The controls so obtained could be stored, and the one "closest" to the state to be controlled could be used to steer the state. Other schemes, such as the construction of switching hypersurfaces or the approximation of cost surfaces by suitable scalar functions of the state as in Reference 20, come immediately to mind as suboptimal-control schemes that might be designed using a high-speed computer and an iterative procedure for computing optimal controls in some design procedure.

2.3 First Iterative Method [19][1]

This method is due to Y. C. Ho and has been improved by P. S. Fancher in Reference 15. Problem Z_5 (Definition 1.14) with (1) $\dot{\mathbf{x}}(t) = \mathbf{A}(t)\mathbf{x}(t) + \mathbf{b}(t)u(t)$, (2) $|u| \leqslant 1$, (3) $L = 0$, (4) $F = \|\mathbf{x}(T)\|$, and (5) $h = 1$ is solved successively with increasing times T until $\mathbf{x}(T) = 0$ results. The last problem in this sequence is one whose solution is also that of the time-optimal regulator problem. The fixed-time problem is, in turn, solved by a method of successive approximations to the optimal-control sequence.

A less general statement of the problem will no doubt be appreciated.

> Let (1) $\dot{\mathbf{x}}(t) = \mathbf{A}(t)\mathbf{x}(t) + \mathbf{b}(t)u(t)$, (2) $\mathbf{x}(t_0) = \mathbf{x}_0$, (3) $|u(t)| \leqslant 1$ on $(t_0, T]$, (4) T be fixed, and (5) $J = \|\mathbf{x}(T)\|$. Then the problem is to find the control $u_{(t_0,T]}$ that minimizes J.

If T is large enough, one has $\mathbf{x}^*(T) = \mathbf{0}$, and the least T, say T^*, such that $\mathbf{x}^*(T^*) = \mathbf{0}$, is the minimum time for the unmodified optimal regulator problem. The iterative procedure for approximating Problem Z_1 successively by problem Z_5 is as follows: (1) choose some $T \leqslant T^*$; (2) solve problem Z_5; (3) increase T and repeat until $x(T) = 0$ and $T = T^*$. Ho's method for solving Z_5 by successive approximation is based on the linearity of the plant and on the fact that one can always find a control sequence which is better, in the sense of minimizing $\|\mathbf{x}(T)\|$, than the control sequence $u_{(t_0,T]} = 0$. Let $\mathbf{x}^{(i)}(T)$ denote the terminal state due to a control $u_{(t_0,T]}^{(i)}$ and let $u_{(t_0,T]}^{(0)} = 0$. Then, there are two steps in Ho's iterative procedure for Problem Z_5.

1. Find a $u_{(t_0,T]}^{(1)}$, such that (for the same initial state $\mathbf{x}_0$) one obtains $\mathbf{x}^{(1)}(T)$, such that

$$\|\mathbf{x}^{(1)}(T)\| < \|\mathbf{x}^{(0)}(T)\|.$$

2. Using $u_{(t_0,T]}^{(0)} = 0$, integrate the equations in reverse time from the final state $\mathbf{x}^{(1)}(T)$ to find a new initial state $\mathbf{x}^{(2)}(t_0)$.

---

[1] The ordering of these methods is neither meritorious nor chronological.

Step 1 is then repeated, using $\mathbf{x}^{(2)}(t_0)$ as the initial state and some $u_{(t_0,T]}^{(0)}$. After n repetitions of this process, the nth approximation to the optimal control is, by superposition,

$$u^*_{(t_0,T]} \cong \sum_{i=1}^{n} u_{(t_0,T]}^{(i)}. \tag{2.4}$$

Ho has devised a scheme for choosing $u_{(t_0,T]}^{(i)}$ in which the entire preceding control is stored, and for which the sequence $\|\mathbf{x}_{(T)}^{(i)}\|$ is monotone decreasing.[2]

The storage of the entire preceding control leads either to unnecessary approximation or to storage problems in the computer. Fancher, Reference 15, has eliminated this difficulty, and these results are now summarized. Actually, the results presented in Reference 15 apply to a somewhat more general problem. We will call it Problem $Z\frac{1}{5}$.

We let (1) $\dot{\mathbf{x}}(t) = \mathbf{A}(t)\mathbf{x}(t) + \mathbf{B}(t)\mathbf{u}(t)$, (2) $\mathbf{x}(t_0) = \mathbf{x}_0$, (3) $\mathbf{u}(t) \in \Omega$; Ω convex and compact, (4) $\mathbf{Q}$ be an $n \times n$, positive definite symmetric matrix, (5) $\mathbf{z}$ be a fixed point in R_n, (6) T be the fixed terminal time, and (7) $J = \frac{1}{2}\langle(\mathbf{x}(T) - \mathbf{z}), \mathbf{Q}(\mathbf{x}(T) - \mathbf{z})\rangle$. Then the problem is to find the control that minimizes J.

The generalization here is that (1) the control-constraint set can be any convex and compact subset in R_m, (2) the target can be any point $\mathbf{z}$ in R_n, and (3) the measure of closeness to $\mathbf{z}$ can be varied by a suitable choice of $\mathbf{Q}$.

A digital-computer program has been written in Fortran II for solving this problem when $\Omega = (\mathbf{u}\colon |u_i| \leq 1, i = 1, 2, \ldots, m)$, and this program is presented in Appendix H.

We let (1) the subscript k denote the kth iteration; (2) $\mathbf{x}_k(T)$ be the terminal state due to the application of the control $\mathbf{u}_{k(t_0,T]}$; (3) $\mathbf{c}_k(t) = \mathbf{B}'(t)\phi'(T, t)\mathbf{Q}\mathbf{y}_k$ (see Definition 1.3) ('superscript denotes transpose), where $\mathbf{y}_k = \mathbf{x}_k(T) - \mathbf{z}$; (4) $\bar{\mathbf{u}}_k(t)$ be the control which minimizes $\langle \mathbf{c}_k(t), \mathbf{u}(t)\rangle$; (5) $\bar{\mathbf{x}}_k(T)$ be the terminal state after the application of the control $\bar{\mathbf{u}}_{k(t_0,T]}$, $\bar{\mathbf{y}}_k = \bar{\mathbf{x}}_k(T) - \mathbf{z}$ and $\mathbf{w}_k = \bar{\mathbf{x}}_k(T) - \mathbf{x}_k(T)$; and finally (6) $\lambda_k = \mathrm{sat}(-\langle \mathbf{y}_k, \mathbf{Q}\mathbf{w}_k\rangle / \langle \mathbf{w}_k, \mathbf{Q}\mathbf{w}_k\rangle)$, where $\mathrm{sat}(\gamma) = \gamma$, if $|\gamma| \leq 1$ and $= 1$ if $|\gamma| > 1$. The iterative procedure is then

$$\mathbf{u}_0(t) = \mathbf{0}, \qquad t \in (t_0, T], \tag{2.5}$$

$$\mathbf{u}_{k+1}(t) = (1 - \lambda_k)\mathbf{u}_k(t) + \lambda_k\bar{\mathbf{u}}_k(t), \tag{2.6}$$

until

$$\langle \mathbf{y}_k, \mathbf{Q}\mathbf{w}_k\rangle = 0. \tag{2.7}$$

---

[2] The fact that $\|\mathbf{x}^{(i)}(T)\|$ is monotone decreasing does not imply that $\lim_{i\to\infty} \|\mathbf{x}^{(i)}(T)\| = J^*$.

The step implied by Equation 2.6 is not actually carried out on the computer. The iterative procedure can be programmed as follows.

1. Set $\mathbf{u}_0(t) = \mathbf{0}$, and solve for $\mathbf{x}_0(T) = \boldsymbol{\phi}(T, t_0)\mathbf{x}(t_0)$ (assumed $\neq \mathbf{0}$).

2. Find $\bar{\mathbf{x}}_0(T) = \boldsymbol{\phi}(T, t_0)(\mathbf{x}(t_0) + \int_{t_0}^{T} \phi(t_0, s)B(s)\bar{\mathbf{u}}_0(s)\, ds)$, where $\bar{\mathbf{u}}_0(t)$ minimizes $\langle \mathbf{c}_0(t), \mathbf{u}(t)\rangle = \langle \mathbf{B}'(t)\boldsymbol{\phi}'(T, t)(\mathbf{x}_0(T) - \mathbf{z}), \mathbf{u}(t)\rangle$.

3. Choose $\lambda_1 = \text{sat}(-\langle \mathbf{y}_0, \mathbf{Q}\mathbf{w}_0\rangle / \langle \mathbf{w}_0, \mathbf{Q}\mathbf{w}_0\rangle)$, and, since the plant is linear, we have

$$\mathbf{x}_1(T) = (1 - \lambda_0)\mathbf{x}_0(T) + \boldsymbol{\lambda}_0\bar{\mathbf{x}}_0(T).$$

Steps 2 and 3 are repeated until Equation 2.7 is satisfied.

This technique is illustrated in Figure 2.1, where a hypothetical set of reachable states is shown. The initial guess $\mathbf{u}_{0(t_0,T]} = \mathbf{0}$ results in $\mathbf{x}_0(T)$ in the interior of the set of reachable states R from $\mathbf{x}(t_0)$. The vector $\mathbf{y}_0 = \mathbf{x}_0(T) - \mathbf{z}$ is used to give the direction of $\mathbf{P}_0(T)$ [that is, the final value of the costate which is used to generate an extremal control terminating at $\bar{\mathbf{x}}_0(T)$]. Because it is the terminal state after the application of an extremal control, $\bar{\mathbf{x}}_0(T)$ is on the boundary of R at the point where the support hyperplane to R (see Figure 2.1) has a normal vector, inward to R, whose direction is that of $\mathbf{p}_0(T) = \mathbf{y}_0$. The

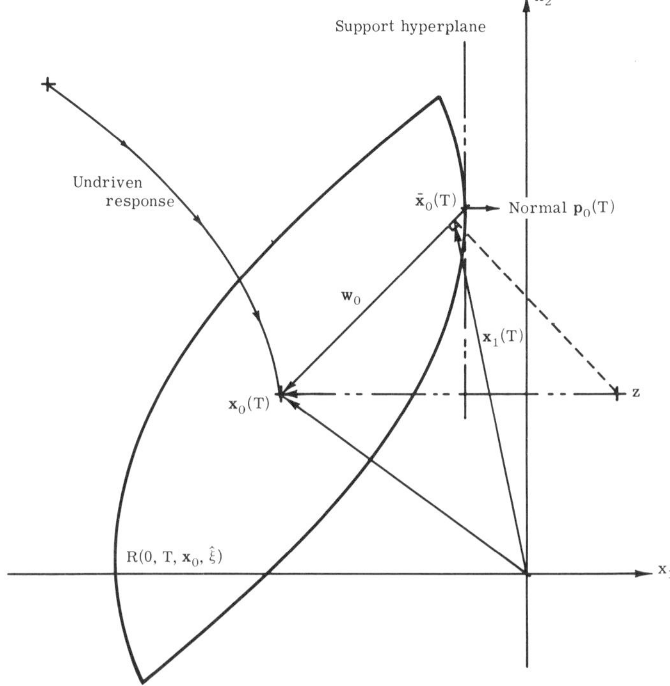

Figure 2.1 Set of reachable states.

vector $\mathbf{w}_0$ and the scalar λ_0 are defined such that $\mathbf{x}_1(T)$ will lie on the line segment joining $\bar{\mathbf{x}}_0(T)$ and $\mathbf{x}_0(T)$ at the point which is closest to $\mathbf{z}$. In Figure 2.1, we assume that $\mathbf{Q} = \mathbf{I}$, and hence the line segment joining $\mathbf{z}$ and $\mathbf{x}_1(T)$ is perpendicular to the line segment joining $\bar{\mathbf{x}}_0(T)$ and $\mathbf{x}_0(T)$.

This procedure is closely related to the necessary conditions of the Minimum Principle (Theorem 1.2). In this case, the Hamiltonian is

$$H = p_0 + \langle \mathbf{p}(t), \mathbf{A}(t)\mathbf{x}(t)\rangle + \langle \mathbf{p}(t), \mathbf{B}(t)\mathbf{u}(t)\rangle, \tag{2.8}$$

and NC11 implies that

$$\mathbf{p}(T) = F_{\mathbf{x}}(\mathbf{x}(T)) = \mathbf{Q}(\mathbf{x}(T) - \mathbf{z}). \tag{2.9}$$

Choosing $\mathbf{c}_k(t) = \mathbf{B}'(t)\phi'(T, t)\mathbf{Q}(\mathbf{x}_k(T) - \mathbf{z})$ and $\mathbf{u}_k(t)$ to minimize Equation 2.8 is the same as simply using $\mathbf{x}_k(T) - \mathbf{z}$ as the current estimate of the final value of the optimal costate vector. It is known [34] that given a vector $\mathbf{p}(T)$, the choice of $\bar{\mathbf{u}}_{(t_0,T]}$ such that $\bar{\mathbf{u}}(t)$ minimizes the Hamiltonian (Equation 2.8) results in a trajectory with a $\bar{\mathbf{x}}(T)$ at the point on the boundary of the set of reachable states R whose support hyperplane, at that point, has a normal, directed into R, with the direction of $\mathbf{p}(T)$. Then, at each step one finds the extremal control associated with the final costate vector whose direction is that of the vector $\mathbf{Q}(\mathbf{x}(T) - \mathbf{z})$. Step 3 corresponds to choosing the kth terminal state at that point on the line segment joining $\mathbf{x}_{k-1}(T)$ to $\bar{\mathbf{x}}_{k-1}(T)$ which is closest to $\mathbf{z}$, the corresponding kth estimate to the optimal control being

$$\mathbf{u}_k(t) = (1 - \lambda_k)\mathbf{u}_k(t) + \lambda_k\bar{\mathbf{u}}_k(t).$$

It is assumed that $\bar{\mathbf{u}}_k(t)$ is well defined (that is, that $\langle \mathbf{c}_k(t), \mathbf{u}(t)\rangle$ has a unique minimum for all but a finite number of times t in $(t_0, T]$). In other words, it is assumed that the problem is normal (Definition 1.10). The minimum of $\langle \mathbf{c}_k(t), \mathbf{u}(t)\rangle$ lies on $\partial\Omega$, the boundary of Ω, at the point (or points) where the normal to the support hyperplane of Ω has the direction of $-\mathbf{c}_k(t)$ (Appendix C). These points are unique unless $\mathbf{c}_k(t)$ is perpendicular to a line segment of $\partial\Omega$. It follows that the problem is normal if $\mathbf{c}_k(t)$ can be perpendicular to a line segment of $\partial\Omega$ at most a finite number of times in any interval $(t_0, T]$. When Ω is the unit hypercube (Problem Z_1 Definition 1.11), and if $\mathbf{A}(t)$ and $\mathbf{B}(t)$ are constant, one requires that all of the controllability matrices (Definition 1.8) be nonsingular (Appendix C).

Lemma 2.1

If $\mathbf{y}_k = \mathbf{x}_k(T) - \mathbf{z}$ and $\mathbf{w}_k = \bar{\mathbf{x}}_k(T) - \mathbf{x}_k(T)$, then

$$\langle \mathbf{y}_k, \mathbf{Q}\mathbf{w}_k\rangle \leq 0, \tag{2.10}$$

and the equality holds in Equation 2.10 only if $\bar{\mathbf{u}}_k(t) = \mathbf{u}_k(T)$ almost everywhere in $(t_0, T]$.

Proof:

$$\mathbf{x}_k(T) = \phi(T, t_0)(\mathbf{x}(t_0) + \int_{t_0}^{T} \phi(t_0, t)\mathbf{B}(t)\mathbf{u}_k(t)\, dt), \tag{2.11}$$

and

$$\bar{\mathbf{x}}_k(T) = \phi(T, t_0)(\mathbf{x}(t_0) + \int_{t_0}^{T} \phi(t_0, t)\mathbf{B}(t)\bar{\mathbf{u}}_k(t)\, dt), \tag{2.12}$$

whence we have

$$\mathbf{w}_k = \bar{\mathbf{x}}_k(T) - \mathbf{x}_k(T) = \int_{t_0}^{T} \phi(T, t)\mathbf{B}(t)(\bar{\mathbf{u}}_k(t) - \mathbf{u}_k(t))dt \tag{2.13}$$

and

$$\langle \mathbf{y}_k, \mathbf{Q}\mathbf{w}_k \rangle = \int_{t_0}^{T} \langle \mathbf{y}_k, \mathbf{Q}\phi(T, t)\mathbf{B}(t)(\bar{\mathbf{u}}_k(t) - \mathbf{u}_k(t))\rangle dt; \tag{2.14}$$

that is,

$$\langle \mathbf{y}_k, \mathbf{Q}\mathbf{w}_k \rangle = \int_{t_0}^{T} \langle \mathbf{c}_k(t), (\bar{\mathbf{u}}_k(t) - \mathbf{u}_k(t))\rangle dt. \tag{2.15}$$

But, by definition, $\bar{\mathbf{u}}_k(t)$ minimizes $\langle \mathbf{c}_k(t), \mathbf{u}(t)\rangle$ uniquely almost everywhere. It follows that $\langle \mathbf{c}_k(t), \bar{\mathbf{u}}_k(t) - \mathbf{u}_k(t)\rangle \leqslant 0$ on $(t_0, T]$ with the strict inequality holding almost everywhere unless $\bar{\mathbf{u}}_k(t) = \mathbf{u}_k(t)$. Hence, $\langle \mathbf{y}_k, \mathbf{Q}\mathbf{w}_k\rangle < 0$, unless $\bar{\mathbf{u}}_k(t) = \mathbf{u}_k(t)$ almost everywhere. Q.E.D.

Lemma 2.2

If Problem Z_5^1 is normal, the extremal controls are unique.

Proof:

Suppose the contrary, that is, assume that there are two unequal extremal controls $\mathbf{u}_1(t_0, T]$ and $\mathbf{u}_2(t_0, T]$, both of which steer the state from $\mathbf{x}_0$ to the same state $\mathbf{x}(T)$. By equating the two corresponding solutions to the state-vector differential equation, it follows that

$$\int_{t_0}^{T} \phi(t_0, t)\mathbf{B}(t)(\mathbf{u}_1(t) - \mathbf{u}_2(t))\, dt = 0, \tag{2.16}$$

from which, by multiplying from the left with $\phi'(T, t_0)\mathbf{Q}(\mathbf{x}(T) - \mathbf{z})$, we get

$$\int_{t_0}^{T} \langle(\mathbf{x}(T) - \mathbf{z}), \mathbf{Q}\phi(T, t)\mathbf{B}(t)(\mathbf{u}_1(t) - \mathbf{u}_2(t))\rangle dt = 0; \tag{2.17}$$

that is,

$$\int_{t_0}^{T} \langle \mathbf{c}_k(t), (\mathbf{u}_1(t) - \mathbf{u}_2(t))\rangle dt = 0. \tag{2.18}$$

But $\mathbf{u}_1$ and $\mathbf{u}_2$ are extremal controls, both of which minimize (uniquely) $\langle \mathbf{c}_k(t), \mathbf{u}(t) \rangle$ almost everywhere. It follows from Lemma 2.1 that $\mathbf{u}_1(t) = \mathbf{u}_2(t)$ almost everywhere. Q.E.D.

Theorem 2.1

If Problem Z_5^1 is normal, the iterative procedure is such that the sequence $\mathbf{u}_{k(t_0,T]}$, $k = 0, 1, \ldots,$ converges to $\overset{*}{\mathbf{u}}_{(t_0,T]}$, the control that minimizes $\langle \mathbf{y}, \mathbf{Q}\mathbf{y} \rangle$. Further, the sequence $\langle \mathbf{y}_k, \mathbf{Q}\mathbf{y}_k \rangle$, $k = 1, 2, \ldots,$ is monotone decreasing.

Proof:

Since

$$\mathbf{u}_{k+1}(t) = (1 - \lambda_k)\mathbf{u}_k(t) + \lambda_k \bar{\mathbf{u}}_k(t), \tag{2.19}$$

$$\mathbf{x}_{k+1}(T) = (1 - \lambda_k)\mathbf{x}_k(T) + \lambda_k \bar{\mathbf{x}}_k(T), \tag{2.20}$$

and

$$\mathbf{y}_{k+1} = \mathbf{y}_k + \lambda_k \mathbf{w}_k, \tag{2.21}$$

we have

$$\langle \mathbf{y}_{k+1}, \mathbf{Q}\mathbf{y}_{k+1} \rangle = \langle \mathbf{y}_k, \mathbf{Q}\mathbf{y}_k \rangle + 2\lambda_k \langle \mathbf{y}_k, \mathbf{Q}\mathbf{w}_k \rangle + \lambda_k^2 \langle \mathbf{w}_k, \mathbf{Q}\mathbf{w}_k \rangle, \tag{2.22}$$

and

$$\langle \mathbf{y}_{k+1}, \mathbf{Q}\mathbf{y}_{k+1} \rangle - \langle \mathbf{y}_k, \mathbf{Q}\mathbf{y}_k \rangle < 0, \tag{2.23}$$

and Relation 2.23 is minimized for $0 < \lambda_k \leq 1$ if

$$\lambda_k = \text{sat}(-\langle \mathbf{y}_k, \mathbf{Q}\mathbf{w}_k \rangle / \langle \mathbf{w}_k, \mathbf{Q}\mathbf{w}_k \rangle), \tag{2.24}$$

provided that

$$\langle \mathbf{y}_k, \mathbf{Q}\mathbf{w}_k \rangle \leq 0. \tag{2.25}$$

In other words, the choice for λ_k in Equation 2.24 achieves the best "local" reduction in the terminal cost. It follows from Lemmas 2.1 and 2.2 that the terminal cost is always reduced unless $\mathbf{u}_{k(t_0,T]}$ is the unique optimal control. Q.E.D.

In Theorem 2.1 we assumed that $T \leq T^*$. Suppose that $T > T^*$; How will the iterative procedure behave then? Lemma 2.1 does not require that $T \leq T^*$, and hence $\langle \mathbf{y}_k, \mathbf{Q}\mathbf{w}_k \rangle \leq 0$ in any case. This means (see Theorem 2.1) that $\langle \mathbf{y}_{k+1}, \mathbf{Q}\mathbf{y}_{k+1} \rangle \leq \langle \mathbf{y}_k, \mathbf{Q}\mathbf{y}_k \rangle$. Furthermore, since the sequence $\mathbf{u}_k$, $k = 0, 1, 2, \ldots,$ cannot converge to a time-optimal control (if $T > T^*$, $\mathbf{z}$ is in the interior of the set of states that are reachable from $\mathbf{x}_0$), we have strict inequalities, that is, $\langle \mathbf{y}_k, \mathbf{Q}\mathbf{w}_k \rangle < 0$ and $\langle \mathbf{y}_{k+1}, \mathbf{Q}\mathbf{y}_{k+1} \rangle < \langle \mathbf{y}_k, \mathbf{Q}\mathbf{y}_k \rangle$. It follows that $\lim_{k \to \infty} (\langle \mathbf{y}_k, \mathbf{Q}\mathbf{y}_k \rangle) = 0$, and $\mathbf{u}_{k \to \infty}(t_0, T]$ is not an optimal control. In other words, if we choose $T > T^*$, the iterative procedure of this section will converge to a feasible control for the problem of "hitting" $\mathbf{z}$.

The next question is: Is there a simple way to determine if the given choice for T is greater than or less than T*? There is such a simple test. It can be easily seen that if $T \leqslant T^*$,

$$\operatorname*{Lim}_{k \to \infty} (\mathbf{x}_k(T)) = \lim_{k \to \infty} (\bar{\mathbf{x}}_k(T)),$$

whereas if $T > T^*$, the limits do not coincide. In other words, if we find that the sequence $\langle \mathbf{y}_k, \mathbf{Q}\mathbf{y}_k \rangle$ is approaching zero but that $\mathbf{x}_k(T)$ is not approaching $\bar{\mathbf{x}}_k(T)$, then we conclude that $T > T^*$. This fact is useful if we are using Problem Z_5^1 successively to solve Problem Z_1, since it will then be possible to tell whether the given guess T is larger or smaller than T*, and a method of halving the interval containing T* could be used to find T*.

Since few experimental results are published, it is difficult to compare this procedure with others in a quantitative manner. However, a few comments can be made: (1) The amount of computation required per iteration is modest, since the differential equations are integrated only once. This aspect is important since, from this author's experience, carrying out an integration is by far the most time-consuming portion of an algorithm for solving optimal-control problems on a digital computer. In fact, these problems may be more rapidly solved on a hybrid computer; (2) The ease with which one can carry out an integration will depend on the properties of the set Ω, that is, on the difficulties of computing $\bar{\mathbf{u}}_k(t)$; (3) This solution technique relies heavily on the linearity of the plant (see Equation 2. 20). It is doubtful that the approach will extend nontrivially to the nonlinear case. The reason for this is that the set of reachable states in the nonlinear case is, in general, not a convex set.

2.4 Second Iterative Method [13], [14], [29], [30], [35]

This method is due to Neustadt [29] and [30]. Eaton [13] has extended it to the rendezvous problem. Some computational aspects have been published by Fadden and Gilbert [14] and by Paiewonsky [35].

Theorem 2. 2

If the plant in Problem Z_1 is linear and if the target set is the origin (that is, if $r = 0$) and if the problem is normal, then the time-optimal control that takes the state from some controllable initial state to the origin exists, is unique, and is given by

$$\mathbf{u}^*(t) = -\text{sgn}\{\pi'^*\mathbf{\Phi}(t, t_0)\mathbf{B}(t)\}, \tag{2.26}$$

where

$$\pi^* \neq \mathbf{0},$$

and $\mathbf{\Phi}'(t, t_0)\pi^*$ is the solution of the costate differential equations

$$\dot{\mathbf{p}}^*(t) = -\mathbf{A}'(t)\mathbf{p}^*(t), \qquad \mathbf{p}^*(t_0) = \pi^*. \tag{2.27}$$

The proof of Theorem 2.2 may be found in Reference [3] or [32].

Equation 1.7 admits the general solution

$$\mathbf{x}(t) = \mathbf{\Phi}(t, t_0)\mathbf{x}_0 + \int_{t_0}^{t} \mathbf{\Phi}(t, s)\mathbf{B}(s)\mathbf{u}(s)\, ds, \tag{2.28}$$

and, since $\mathbf{x}(T) = \mathbf{0}$, we have

$$-\mathbf{x}_0 = \int_{t_0}^{T} \mathbf{\Phi}(t_0, s)\mathbf{B}(s)\mathbf{u}(s)\, ds. \tag{2.29}$$

Define a set in R_n,

$$\psi(t_0, T) \equiv \{\mathbf{x} : \mathbf{x} = -\int_{t_0}^{T} \mathbf{\Phi}(t_0, s)\mathbf{B}(s)\mathbf{u}(s)\, ds, \mathbf{u} \in \Omega_a\}. \tag{2.30}$$

Properties of $\psi(t_0, T)$ [27]

1. $\psi(t_0, T)$ is the set of states from which, at t_0, the origin can be reached in time t_1 and with $\mathbf{u} \in \Omega_a$.
2. $\psi(t_0, T)$ is convex and compact.
3. $\psi(t_0, T_1) \subset \psi(t_0, T_2)$ if $T_2 \geqslant T_1$.
4. $\psi(t_0, T)$ grows continuously with T.

Properties 3 and 4 follow easily from Equation 2.29. Let $\mathbf{u}_{(t_0,T_2]} = \mathbf{u}_{(t_0,T_1]} + \mathbf{u}_{(T_1,T_2]}$. From Equation 2.29, a choice of $\mathbf{u}_{(T_1,T_2]} = \mathbf{0}$ implies that $\psi(t_0, T_1) \subset \psi(t_0, T_2)$. A similar argument with $\mathbf{u}_{(T_1,T_2]} \neq 0$ shows Property 4. The convexity of $\psi(t_0, T_1)$ follows directly from the fact that Ω_a is convex and that Equation 2.29 describes a linear mapping from Ω_a into R_n. The proof of the compactness of $\psi(t_0, T)$ can be found in Reference 29 or 24. It is obvious from these properties that the boundary of $\psi(t_0, T)$ contains all the states for which a time-optimal policy is required to reach the origin. Hence the significance of $\psi(t_0, T)$ compact; for if it were not compact, the boundary would not be defined, and the optimal control would not exist. Since the boundary of $\psi(t_0, T)$ is the set of states for which an optimal-control policy is required to take the state to the origin in time T, this boundary is called a minimum-cost surface, or in the case of time-optimal control, a minimum isochrone.

Since $\psi(t_0, T)$ is convex and compact, it has a support hyperplane everywhere. It can be shown that the optimal costate vector $\mathbf{p}^*(t)$ is the outward normal to a support hyperplane of $\psi(t_0, t)$ at $\mathbf{x}^*(t)$.[3] In fact, this is the physical significance of the costate, that it is colinear with the negative direction of the gradient of the minimum cost, whereever the latter exists. It follows, in view of Equations 2. 26 and 2. 27, that choosing a π yields an optimal control for some state $\hat{\mathbf{x}}$ on the boundary of $\psi(t_0, T)$. From Figure 2. 2 it is clear that many π's can yield the same optimal control, and hence correspond to the same $\hat{\mathbf{x}}$.

These properties of $\psi(t_0, T)$ can be used to produce the following iterative procedure for the normal problem.

1. Choose a $T > 0$ and a π such that[4]

$$\langle \pi, \mathbf{x}_0 \rangle > 0. \tag{2. 31}$$

2. Find

$$\hat{\mathbf{x}}(t_0, t_1; \pi) = \int_{t_0}^{T} \Phi(t_0, s)\mathbf{B}(s)\operatorname{sgn}[\pi' \Phi(t_0, s)\mathbf{B}(s)]\, ds. \tag{2. 32}$$

Now, $\hat{\mathbf{x}}(t_0, T; \pi)$ belongs to the boundary of $\psi(t_0, T)$, and, since $\psi(t_0, t_1)$ is convex, all other points in $\psi(t_0, T)$ have a smaller projection on the direction of π than $\hat{\mathbf{x}}$ does (see Figure 2. 2); that is, we have

$$\langle \pi, \hat{\mathbf{x}}(t_0, T; \pi)\rangle > \langle \pi, \mathbf{x}\rangle, \qquad \text{for all } \mathbf{x} \in \psi(t_0, T), \qquad \mathbf{x} \neq \hat{\mathbf{x}}. \tag{2. 33}$$

Further, $\langle \pi, \hat{\mathbf{x}}(t_0, t; \pi)\rangle$ is strictly increasing in t; that is, we have

$$\langle \pi, \hat{\mathbf{x}}(t_0, t; \pi)\rangle = \int_{t_0}^{t} \sum_{\beta=1}^{m} |\langle \pi, \Phi(t_0, s)\mathbf{b}_\beta(s)\rangle|\, ds. \tag{2. 34}$$

3. Find $\hat{t}$, such that

$$g(t_0, \hat{t}, \pi, \mathbf{x}_0) \equiv \langle \pi, \mathbf{x}_0 - \hat{\mathbf{x}}(t_0, \hat{t}, \pi)\rangle = 0. \tag{2. 35}$$

Step 3 can be carried out because Equations 2. 30, 2. 32, 2. 33, and 2. 34 imply that

A. $g(t_0, t, \pi, \mathbf{x}_0)$ is strictly increasing in t, (2. 36)

B. $g(t_0, t^*, \pi, \mathbf{x}_0) < 0, \qquad \pi \neq \pi^*,$ (2. 37)

---

[3] [3] Chapter 6.

[4] Since $\psi(t_0, T)$ is convex and contains the origin and since π^* is outwardly normal to a support hyperplane of $\psi(t_0, T^*)$ at $\mathbf{x}_0$, then Equation 2. 31 is a property of π^*.

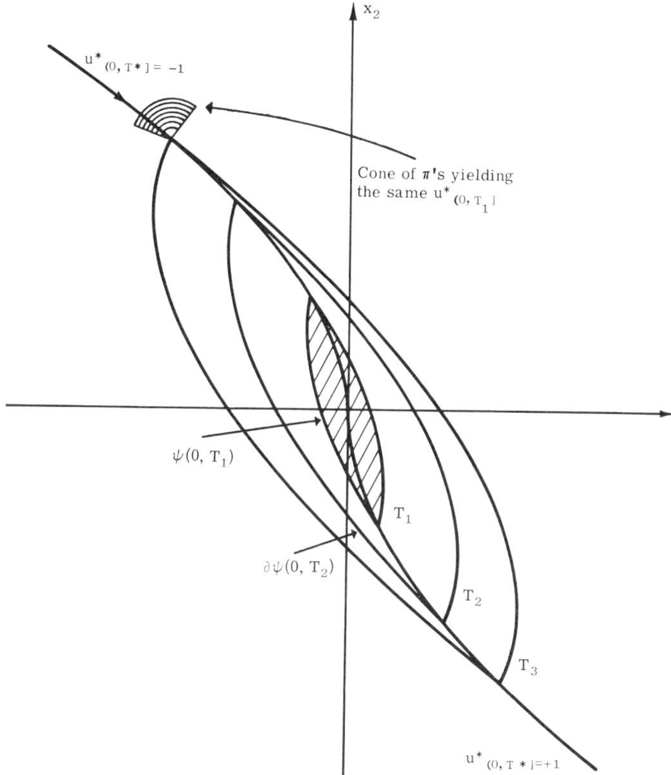

Figure 2. 2 The minimum isochrones and Neustadt's method.

C. $g(t_0, t_0, \pi, \mathbf{x}_0) > 0,$ (2.38)

D. $g(t_0, t^*, \pi^*, \mathbf{x}_0) = 0,$ (2.39)

E. $g(t_0, \hat{t}, \pi, \mathbf{x}_0) = 0$, when the error vector $\mathbf{x}_0 - \hat{\mathbf{x}}(t_0, \hat{t}, \pi)$ lies in the support hyperplane of $\psi(t_0, \hat{t})$ at $\hat{\mathbf{x}}(t_0, \hat{t}, \pi)$. (2.40)

This step implicitly defines a function $\hat{t}(\boldsymbol{\pi}, t_0, \mathbf{x}_0)$ having the property

$$T^* = \max_{\boldsymbol{\pi}} \hat{t}(\pi, t_0, \mathbf{x}_0) = \hat{t}(\boldsymbol{\pi}^*, t_0, \mathbf{x}_0). \tag{2.41}$$

4. Choose a better $\boldsymbol{\pi}$. One way to choose a better π is

$$\boldsymbol{\pi}_{k+1} = \boldsymbol{\pi}_k + \alpha_k(\mathbf{x}_0 - \hat{\mathbf{x}}(\boldsymbol{\pi}_k, \hat{t}_k, t_0)), \qquad \alpha_k > 0. \tag{2.42}$$

Neustadt does not specify how to choose α_k. Other researchers have investigated schemes for making this choice in certain example problems. Fadden and Gilbert [14] assumed that (π^*, T^*) was known and examined a linearized form of Equation 2.41 expanded about that point. Upper bounds on α_k were obtained for stable iteration. There are sets of $\boldsymbol{\pi}$'s on which the linearized version of Equation 2.41 does not exist and near which the upper bound on α_k goes to zero. However, these sets have dimension less than n and are assumed to cause little difficulty.[5] Fadden and Gilbert report experimental results for the triple integrator plant. In these experiments, α_k was chosen by evaluating the upper bounds on the current guess of (π_k, T_k) rather than on (π^*, T^*). The results apparently were favorable when a particular state-space representation was used. When a particular canonical state-space representation was used, the error behavior was very erratic but finally converging. This erratic behavior is probably because (π_k, T_k) was used instead of (π^*, T^*) to evaluate the upper bounds on α_k and/or the fact that the minimum isochrones are extremely flat for the triple integrator when the Jordan state-space representation is used.

Paiewonsky and Neustadt [35] have experimented with a third-order system having $m = 2$ and $u \in \Omega_b$. A gradient type of search similar to Equation 2.41 was used to find the maximum of $\hat{t}(\boldsymbol{\pi}, t_0, \mathbf{x}_0)$. Six different techniques were tested for choosing α_k. Of these, two produced favorable results. One was Powell's (see [35]) method of quadratic approximation, and the other is described as "a variable-gain method using scalar products of successive gradients." It was found that $\hat{t}(\boldsymbol{\pi}, t_0, \mathbf{x}_0)$ is extremely flat. For example, a change in the third significant figure in the evaluation of $\hat{t}(\boldsymbol{\pi}, t_0, \mathbf{x}_0)$ corresponded to a change of three orders of magnitude in the Euclidean norm of the error $(\|\mathbf{x}_0 - \hat{\mathbf{x}}(\pi, t_0, \mathbf{x}_0)\|)$. It is suggested in Reference 35 that this sensitivity requires the use of double precision in the calculation of $\hat{t}(\boldsymbol{\pi}, t_0, \mathbf{x}_0)$.

A final comment is that this procedure is based on the convexity of $\psi(t_0, T)$. Few nonlinear systems that is nonlinear in $\mathbf{x}$ will possess this property, and hence the method will not extend readily to the solution of the time-optimal control of nonlinear plants.

The reader is referred to a discussion of the use of this method on minimum-effort problems in Reference 31.

2.5 Third Iterative Method [33][6]

This method, attributed to N. N. Krasovski and R. V. Gamkrelidze, uses the Minimum Principle to solve Problem Z_1, and, as in Neustadt's

[5] These sets of $\boldsymbol{\pi}$'s also occur in one of the iterative procedures proposed in this monograph.

[6] Pages 111–115.

method, a search is made among the costate initial conditions (the $\{\pi\}$'s) and among the control intervals for the ordered pair (π^*, T^*). Only the basic idea is reported here.

The method is as follows: (1) choose a $T < T^*$, (2) find a $\hat{\pi}$ such that $\hat{\mathbf{x}}(\hat{\pi}, T)$ is colinear with $\mathbf{x}_0$, (3) increase T and repeat until $\hat{\mathbf{x}} = \mathbf{x}_0$. The corresponding $\hat{\pi}$ and T will be optimal.

Gradient techniques for finding $\hat{\pi}$ such that $\hat{\mathbf{x}}(\hat{\pi}, T) = \alpha \mathbf{x}_0, \alpha > 0$, are discussed in Reference 35 along with some experimental results. These results are compared [35] with those found when using Neustadt's method on the same problem. The methods are described as being equally successful.

The basic idea behind this approach is capable of wider application than Neustadt's method, since the linearity of the plant is not explicitly assumed. Hence, if Step 2 can be carried out in a nonlinear case and if certain conditions hold (such as the uniqueness of the extremal controls, for example), then this approach would be applicable.

2.6 Fourth Iterative Method [8], [22], [33], [36], [37]

This is the most formal approach. The application of the Maximum Principle leads to a two-point boundary-value problem. The iterative procedure is to guess the unknown boundary conditions at one point, solve the initial-value problem, and find the error on the conditions at the other point. At the same time, the first-order effects of changing the guessed values are calculated. These first-order effects are then used to find a better guess. The method does not always work, because the first-order effect of varying some of the possible guesses is zero. One purpose of this work is to show how a modification in the formulation of the problem removes this difficulty.

The references cited deal with the time-optimal problem. Though the general method is in no way restricted to this problem, it is reviewed in that light here. The plant studied by Bass [8] is

$$\dot{\mathbf{x}}(t) = \mathbf{f}(\mathbf{x}(t)) + \mathbf{K}(\mathbf{x}(t))\mathbf{u}(t), \qquad (2.43)$$

where $\mathbf{K}(\mathbf{x}(t))$ is a matrix whose elements are functions of the state. Paiewonsky [33] and Knudsen [22] have treated the case

$$\dot{\mathbf{x}}(t) = \mathbf{A}\mathbf{x}(t) + \mathbf{b}u(t). \qquad (2.44)$$

For the sake of variety, consider Equation 2.43. Using the Minimum Principle, we find that the Hamiltonian is

$$H(\mathbf{x}, \mathbf{u}, p) = p_0 + \langle \mathbf{p}, \mathbf{f}(\mathbf{x}) \rangle + \sum_{i=1}^{m} \langle \mathbf{p}, \mathbf{k}_i(\mathbf{x}) \rangle u_i, \qquad (2.45)$$

where $\mathbf{k}_i(\mathbf{x})$, $i = 1, \ldots, m$ is the ith column vector of $\mathbf{K}(\mathbf{x})$. If the problem is normal—actually little is known in detail about the normality of this problem unless $\mathbf{f}(\mathbf{x})$ is linear in $\mathbf{x}$ and $\mathbf{K}(\mathbf{x}(t)) = \mathbf{K}(t)$—then Equation 2.45 implies a control law:

$$\mathbf{u}^*(t) = -\operatorname{sgn}\{\mathbf{K}'(\mathbf{x}^*(t))\mathbf{p}^*(t)\}, \tag{2.46}$$

where

$$\dot{\mathbf{p}}^*(t) = -\mathbf{f}_{\mathbf{x}}(\mathbf{x}^*(t))\mathbf{p}^*(t) - \sum_{i=1}^{m} \mathbf{k}_{i\mathbf{x}}(\mathbf{x}^*(t))\mathbf{p}^*(t)u_i{}^*(t). \tag{2.47}$$

Equations 2.43, 2.46 and 2.47 define a two-point boundary-value problem that is summarized as follows:

$$\dot{\mathbf{x}}^*(t) = \mathbf{f}(\mathbf{x}^*(t)) - \mathbf{K}(\mathbf{x}^*(t)) \operatorname{sgn}\{\mathbf{K}'(\mathbf{x}^*(t))\mathbf{p}^*(t)\}, \tag{2.48}$$

$$\dot{\mathbf{p}}^*(t) = -\mathbf{f}_{\mathbf{x}}(\mathbf{x}^*(t))\mathbf{p}^*(t) + \sum_{i=1}^{m} \mathbf{k}_{i\mathbf{x}}(\mathbf{x}^*(t))\mathbf{p}^*(t) \operatorname{sgn}\{\langle \mathbf{k}_i(\mathbf{x}^*(t)), \mathbf{p}^*(t)\rangle\}, \tag{2.49}$$

$$\mathbf{x}^*(0) = \mathbf{x}_0, \qquad \mathbf{x}^*(T^*) = \mathbf{0}, \qquad T^* \text{ is unknown.} \tag{2.50}$$

If the optimal control exists, this two-point boundary-value problem will have a solution. It can be seen immediately that this problem will have an infinity of solutions unless $\|\mathbf{p}\|$ is constrained. This is because Equations 2.48 and 2.49 are invariant under multiplication of $\mathbf{p}$ by a positive scalar. Let the constraint on $\mathbf{p}$ be

$$\|\mathbf{p}^*(T^*)\| = 1. \tag{2.51}$$

The solution of Equations 2.48 through 2.51 has been attempted using both the gradient method [33] and a modified form of Newton's method [22]. These procedures can be described as follows: (1) choose an ordered pair (arbitrarily or otherwise) $(\boldsymbol{\pi}_k, T_k)$, $T_k > 0$, and $\|\boldsymbol{\pi}_k\| = 1$; (2) Equations 2.48 and 2.49 with $\mathbf{x}^*(T_k) = \mathbf{0}$ and $\mathbf{p}^*(T_k) = \boldsymbol{\pi}_k$ define an initial-value problem that is solved for an $\hat{\mathbf{x}}(\boldsymbol{\pi}_k, T_k)$. This process of choosing an ordered pair $(\boldsymbol{\pi}_k, T_k)$ defines a relation between the state space and the product space of the set of all $\boldsymbol{\pi}$'s such that $\|\boldsymbol{\pi}\| = 1$ and $T > 0$. An error vector can be defined as

$$\mathbf{e}(\boldsymbol{\pi}_k, T_k) = \mathbf{x}_0 - \hat{\mathbf{x}}(\boldsymbol{\pi}_k, T_k). \tag{2.52}$$

Equations 2.52 and

$$\|\boldsymbol{\pi}_k\| = 1 \tag{2.53}$$

are considered as a set of n + 1 equations in n + 1 unknowns and may be written in the form

$$\mathbf{F}(\boldsymbol{\pi}, T; \mathbf{x}_0) = \mathbf{0}, \tag{2.54}$$

which, by assumption (existence of the optimal control), has at least one solution (π^*, T^*). It follows immediately that the solution to Equation 2.54 may not be unique because (see Figure 2.2) many π's can generate the same sequence $\mathbf{u}_{(0,T]}$. Let the ordered pairs (π_k, T_k) and (π_k, T_{k+1}) denote two successive guesses to the solution of Equation 2.54, and let

$$\Delta\pi_k = \pi_{k+1} - \pi_k \tag{2.55}$$

and

$$\Delta T_k = T_{k+1} - T_k. \tag{2.56}$$

If $\mathbf{F}$ is differentiable in the sense of Frechet (see Appendix B), then one can find the first-order effects of varying π and T. It is the manner in which these first-order effects are used to establish a new guess which differentiates the gradient method from Newton's method.

Newton's method. If the $(n + 1) \times (n + 1)$ matrix $[\mathbf{F}_\pi(\pi_k, T_k) \mid \mathbf{F}_T(\pi_k, T_k)]$ exists and is nonsingular, an iterative procedure can be built around the recursion formula

$$\begin{bmatrix} \Delta\pi_k \\ --- \\ \Delta T_k \end{bmatrix} = -\alpha_k \left[\mathbf{F}_\pi(\pi_k, T_k) \mid \mathbf{F}_T(\pi_k, T_k) \right]^{-1} \mathbf{F}(\pi_k, T_k). \tag{2.57}$$

This formula with $\alpha_k = 1$ is due to Sir Isaac Newton.[7] Knudsen [22] used a formula similar to Equation 2.57, where instead of using Equation 2.53 as a constraint, he used a dummy row to replace the (n + 1)st row in $[\mathbf{F}_\pi \mid \mathbf{F}_T]$. This dummy row was chosen to be linearly independent of the other n rows. Knudsen used the scalar α_k to control the

$$\left\| \begin{pmatrix} \Delta\pi_k \\ \hline \Delta T_k \end{pmatrix} \right\|.$$

The difficulties associated with this approach are as follows:

1. The derivatives $\mathbf{F}_\pi(\pi, T)$ and $\mathbf{F}_T(\pi, T)$ do not exist everywhere. (This fact is discussed in detail in the sequel.)
2. If the number of switchings is less than n, the rank of $[\mathbf{F}_\pi \mid \mathbf{F}_T]$ is at most one plus the number of switchings. (This rules out the use of this method when the number of switchings is less than the order of the plant.)[8]

---

[7] Though Sir Isaac Newton did not treat the optimal-control problem.

[8] If the eigenvalues of a linear-constant plant are real, the time-optimal control sequence has at most $n - 1$ switchings [3].

3. If the number of switchings is greater than n − 1, it is not clear under what conditions this will ensure the existence of $[\mathbf{F}_\pi \mid \mathrm{F}_\mathrm{T}]^{-1}$.

The advantages of this approach, if it were possible, are as follows:

1. If the guess (π_k, T_k) is close enough to the solution, the procedure will converge rapidly ([7] or [40]) to the solution (π^*, T^*) with $\alpha_k = 1$.

2. If $\alpha_k \neq 1$ is used to control

   $$\left\| \begin{pmatrix} \Delta\pi_k \\ \hline \Delta T_k \end{pmatrix} \right\|$$

   as in [22], the method should converge rapidly because all the parameters are adjusted simultaneously.

<u>The gradient method.</u> To each guess (π_k, T_k), there is associated an error

$$\mathbf{e}(\pi_k, T_k; x_0) = \mathbf{F}(\pi_k, T_k; x_0). \tag{2.58}$$

The idea of the gradient method is to search a scalar-valued function of (π, T) for a unique minimum (or maximum) occurring at (π^*, T^*). The Euclidean norm of $\mathbf{e}(\pi_k, T_k; x_0)$, $\|\mathbf{e}(\pi_k, T_k; \mathbf{x}_0)\|$ is a candidate for such a function, since it is zero if and only if (π, T) satisfies Equation 2.54, and otherwise it is positive. However, it will not have a unique minimum for all $\mathbf{x}_0$, because π^* may not be unique. (No matter, any π^* will suffice!) The recursion formula for the gradient method is

$$\begin{bmatrix} \Delta\pi_k \\ \hdashline \Delta T_k \end{bmatrix} = -\alpha_k \left[\mathbf{F}_\pi(\pi_k, T_k) \mid \mathbf{F}_\mathrm{T}(\pi_k, T_k)\right]' \quad \mathbf{F}(\pi_k, T_k; \mathbf{x}_0), \tag{2.59}$$

where $\alpha_k > 0$ is chosen to control

$$\left\| \begin{pmatrix} \Delta\pi_k \\ \hline \Delta T_k \end{pmatrix} \right\|.$$

(Usually a search is made over some interval $(0, \rho_k]$ for an α_k that minimizes $\|\mathrm{e}(\pi_{k+1}(\alpha_k), T_{k+1}(\alpha_k)\|)$.

1. This approach has the same difficulties associated with it as Newton's method. That is, if $[\mathbf{F}_\pi \mid \mathbf{F}_\mathrm{T}]$ is singular, then its transpose has a null space. If $\mathbf{F}(\pi_k, T_k)$ lies in the null space of $[\mathbf{F}_\pi(\pi_k, T_k)\ \mathbf{F}_\mathrm{T}(\pi_k, T_k)]'$, then the procedure "hangs up" because no correction can be made. Paiewonsky [33] found large regions in the set of allowable π's for which this occurred.

2. If $[\mathbf{F}_\pi \mid \mathbf{F}_\mathrm{T}]$ is nonsingular and the guess is close enough, Newton's method will be faster than the gradient techniques and hence will require less computation.

3. The nonexistence of $\mathbf{F}_\pi$ and $\mathbf{F}_T$ cause problems which also beset the practical application of the use of Neustadt's method (see [14] and [35]).

It will be shown in this monograph that the problem concerning the nonexistence of $[\mathbf{F}_\pi \vdots \mathbf{F}_T]^{-1}$ (the difficulty found in applying Newton's method) can be eliminated by approximating the optimal-regulator problem with the modified optimal-regulator problem. It will be argued that the modified problem is more meaningful in the engineering sense. There are three reasons for wanting to remove the difficulties from this type of procedure. First, this approach does not rely explicitly on the linearity of the plant, and secondly it is the only approach, of those discussed, in which all of the unknown parameters, the time and the costate boundary conditions, are adjusted simultaneously. It is felt that simultaneous adjustment of these parameters will result in a better iterative scheme. Finally, if the guess is close enough, Newton's method ($\alpha_k = 1$) will apply. This is somewhat satisfying, since an upper bound on the rate of convergence of this method exists and is rapid.

2.7 Engineering Significance

One seeks to modify the optimal-regulator problem in a manner which both makes engineering sense and has the property that an approach similar to that discussed in Section 2.6 can be used to provide a rapidly converging iterative procedure. It is claimed here that the modified optimal-regulator problem (that is, the problem of hitting a small target set containing the origin) is more realistic from the engineering point of view in many cases because of the following reasons:

1. The presence of noise in both the control and in the measurement of the state makes it impossible to control exactly to a point.

2. The mathematical models of many plants are not valid for small norms of the state vector, and hence it is illogical to use an optimal-control law, based on that model, within a small hypersphere about the origin. For example, consider a system having stiction.

3. In any iterative computational scheme, the round-off error requires one to terminate the iteration when the error lies within a prescribed tolerance (sphere).

4. In any case, a policy of using optimal control to the hypersphere followed by a linear feedback control to the origin will be a good suboptimal policy. Such a reversion to linear

feedback control is necessary even in the direct methods in order to eliminate the inevitable limit cycle caused by imperfect "optimal" equipment.[9]

2.8 A General Approach for Solving Problems

The Minimum-Principle point of view is taken, and an equivalent two-point boundary-value problem is constructed. This two-point boundary-value problem is then regarded as defining a function **M** that maps the boundary conditions on the differential equations into the state space. The solution of the optimal-control problem then becomes equivalent to the problem of finding the inverse of **M**. This idea of relating the modified optimal-regulator problem to the problem of inverting a function **M** can be discussed in somewhat general terms. Consider a modified optimal-regulator problem Z, with

$$\dot{\mathbf{x}}(t) = \mathbf{f}(\mathbf{x}(t), \mathbf{u}(t), t), \tag{2.60}$$

$$J = \int_{t_0}^{t_1} L(\mathbf{x}(\tau), \mathbf{u}(\tau), \tau)\, d\tau, \tag{2.61}$$

and assume that (1) $\mathbf{f}(\mathbf{x}, \mathbf{u}, t)$ and $L(\mathbf{x}, \mathbf{u}, t)$ satisfy the conditions of the Maximum Principle (Theorem 1.2), (2) the Problem Z is normal (Definition 1.10), and (3) an optimal control exists for initial conditions $\mathbf{x}_0$ in the domain of controllability C. Assumption 2 means that, given $\mathbf{p}^*_{(t_0,T]}$, $\mathbf{x}^*_{(t_0,T]}$, and p_0^*, we have

$$\mathbf{u}^*(t) = \mathbf{W}(\mathbf{x}^*(t), \mathbf{p}^*(t), p_0^*, t) \qquad (2.61)$$

almost everywhere on $(t_0, T]$. In this hypothetical problem, two situations can occur: (1) either $\mathbf{x}^*(T^*) \in \partial S$, the boundary of S, or (2) $\mathbf{x}^*(T^*) \in S/\partial S$,[10] the interior of S. The set ∂S is an $(n-1)$-fold. Then, if the first situation holds, we have

$$\mathbf{p}^*(T^*) = \alpha \mathbf{x}^*(T^*), \qquad \alpha > 0 \tag{2.62}$$

from NC10 of Theorem 1.2. On the other hand, if the second situation holds, we have the free-end-point problem, and we see that

$$\mathbf{p}^*(T^*) = \mathbf{0} \tag{2.63}$$

from NC11. In both cases, we have

$$\dot{\mathbf{p}}^*(t) = -p_0^* \mathbf{L}_{\mathbf{x}}(\mathbf{x}(t), \mathbf{u}(t), t) - \mathbf{f}'_{\mathbf{x}}(\mathbf{x}(t), \mathbf{u}(t), t)\mathbf{p}(t) \tag{2.64}$$

---

[9] These limit cycles, sometimes referred to as "chattering," can be harmful to relays, etc., though in some cases, such as with positional servomechanisms, they are useful in overcoming stiction.

[10] $S/\partial S$ means the interior of S; see Appendix A.

from NC4 and Equations 1.17, 2.60 and 2.61. Equations 2.60, 2.62, and 2.64, along with the boundary conditions, define the two-point boundary-value problem. Suppose we choose the following:

1. some $T > t_0$, that is, $T \in R_1^+$,
2. some $p_0 \geqslant 0$, that is, $p_0 \in R_1^+$,
3. some n vector $\mathbf{v}$ in S,
4. some $\alpha > 0$, that is, $\alpha \in R_1^+$,
5. if $\mathbf{v} \in \partial S$, $\mathbf{p}^*(T^*) = \alpha\mathbf{v}$ and if $\mathbf{v} \in S/\partial S$, $\mathbf{p}^*(T^*) = 0$

Once a choice is made, the problem becomes an initial-value problem, that is, we let

$$\tau = T + t_0 - t, \tag{2.65}$$

and then we have

$$\dot{\mathbf{x}}(\tau) = -f(\mathbf{x}(\tau), \mathbf{u}(\tau), \tau), \qquad \mathbf{x}(\tau = t_0) = \mathbf{v}, \tag{2.66}$$

$$\dot{\mathbf{p}}(\tau) = + p_0 L_{\mathbf{x}}(\mathbf{x}(\tau), \mathbf{u}(t), \tau) + f'_{\mathbf{x}}(\mathbf{x}(\tau), \tau)\mathbf{p}(\tau), \tag{2.67}$$

$$\mathbf{p}(\tau = t_0) = \mathbf{0} \quad \text{or } \alpha\mathbf{v}, \tag{2.68}$$

$$\mathbf{u}(\tau) = \mathbf{W}(\mathbf{x}(\tau), \mathbf{p}(\tau), p_0, \tau). \tag{2.69}$$

This initial-value problem certainly defines a relation (in an analysis sense) between the set of all ordered choices $(\mathbf{v}, T, p_0, \alpha)$, for which Equation 2.69 is well defined, and a set of vectors $\mathbf{x}$ in R_n. Obviously, if $\mathbf{u}(\tau)$ is well defined, this relation will be a function if and only if unique solutions to the differential Equations 2.66 and 2.67 exist[11] for all choices $(\mathbf{v}, T, p_0, \alpha)$. Assuming that this is true, we have

$$\mathbf{x} = \mathbf{M}(\mathbf{v}, T, p_0, \alpha; t_0). \tag{2.70}$$

In other words, **M** is a function with a domain that is defined to be the set of all $(\mathbf{v}, T, p_0, \alpha)$, for which Equation 2.69 makes sense. From the point of view of the optimal-regulator problem, we are only interested in $\mathbf{v} \in S$. Further, only choices $(\mathbf{v}, T, p_0, \alpha)$ that correspond to an optimal trajectory from some state in C/S to S are of interest.

Definition 2.1 (D)

Let $D \subset S \times R_1^+ \times R_1^+ \times R_1^+$ be such that if $(\mathbf{v}, T, p_0, \alpha) \in D$, then $(\mathbf{v}, T, p_0\alpha)$ corresponds (in the sense of generating an optimal control) to some state in C/S.

---

[11] A basic existence theorem for vector differential equations is given in Reference 3.

Theorem 2.3

Let $\mathbf{M}$ be a function defined on a set D and taking on values in C/S and let $\mathbf{M}$[D] be onto C/S. Then there exists an inverse function $\mathbf{M}^{-1}$ from C/S onto D if and only if $\mathbf{M}$ is one-to-one on D.

Theorem 2.3 is well known and is discussed in Appendix A.

We intend to use an iterative technique similar to that described in Section 2.6 to find $(\mathbf{v}^*, T^*, p_0^*, \alpha^*) = \mathbf{M}^{-1}(\mathbf{x}_0)$. It was shown in Section 2.6 that when $\mathbf{M}$ is many-to-one on D (that is, many π's yield the same optimal control sequence), these methods fail. It will be shown that $\mathbf{M}$ is one-to-one on D in Problems Z_1 through Z_4, inclusive.

Some speculation can be made as to the conditions which imply that $\mathbf{M}$ is one-to-one. From Definition 3.1, we can see that the dimensionality of D is $n + 3$, whereas the range of $\mathbf{M}$ has dimension n. It follows that $\mathbf{M}$ will not be one-to-one on D unless we can discard some of the arguments of $\mathbf{M}$. This is the case in the problems treated here. For example, in Problem Z_1, $\mathbf{M}$ is independent of p_0 and α, and $\mathbf{v}^*$ is constrained to lie on ∂S. This means that D is a set of ordered pairs $(\mathbf{v}, T)$ with $\mathbf{v} \in \partial S$ and $T \in R_1^+$, and thus D has dimension n.

The purpose of this section was to formally establish the point of view taken in the treatment of Problems Z_1 through Z_4 in a somewhat general way. The remainder of the monograph is more specific and treats Problems Z_5, Z_1, Z_2, Z_3, and Z_4 in that order.

3. The Terminal-Cost Problem

3.1 Introduction

A quite general problem (Problem Z_5) is treated in this chapter. The problem is restated formally in Section 3.2, and several lemmas pertaining to its solution are given in Section 3.3. The iterative procedure is stated in Section 3.4, and its convergence is then proved in Section 3.5. Finally, an extension to the case where the control is unconstrained is made in Section 3.6, and the results of the chapter are summarized in Section 3.7.

3.2 The Fixed-Time, Terminal-Cost, Effort-Constrained Problem (Z_5)

Here we restate Problem Z_5 for the convenience of the reader who will, no doubt, note the similarity to the problem treated in Section 2.3.

Given: (1) a fixed interval in time $(t_0, T]$, that is, $(t{:}t_0 < t \leq T)$; (2) a linear plant, that is,

$$\dot{\mathbf{x}}(t) = \mathbf{A}(t)\mathbf{x}(t) + \mathbf{B}(t)\,\mathbf{u}(t), \tag{3.1}$$

where $\mathbf{A}(t)$ is an $n \times n$ matrix and $\mathbf{B}(t)$ is an $n \times m$ matrix, and both matrices are assumed to be continuous in t; (3) an initial state $\mathbf{x}(t_0) = \mathbf{x}_0$; (4) a convex, compact control constraint set $\Omega \subset R_m$; (5) a target point $z \in R_n$; (6) a monotone increasing, convex down function g from R_1 into R_1; (7) a functional $\xi(\mathbf{u}_{(t_0, T]}) = \int_{t_0}^{T} g(\|\mathbf{u}(s)\|)ds$, where $\|\mathbf{u}\|$ means norm of $\mathbf{u}$ (any norm may be used); (8) a number $\hat{\xi}$; and (9) a positive definite constant, symmetric $n \times n$ matrix $\mathbf{Q}$.

Find: the control $\mathbf{u}_{(t_0,T]}$ such that $\mathbf{u}(t) \in \Omega$ for t in $(t_0, T]$, $\xi(\mathbf{u}_{(t_0,T]}) \leqslant \hat{\xi}$, and $\langle(\mathbf{x}(T) - \mathbf{z}), \mathbf{Q}(\mathbf{x}(T) - \mathbf{z})\rangle$ is minimized.

This problem is called "the fixed-interval, free-end-point, terminal-cost, effort-constrained, optimal-control problem." In plainer language, it is the problem of steering the state $\mathbf{x}$, from a given initial state $\mathbf{x}_0$ to a point which is as close as possible to a specified point $\mathbf{z}$, within the confines of specified constraints on both the control vector (that is, Ω) and the total control effort $\xi(\mathbf{u}_{(t_0,T]})$. The term "effort" has been coined by Neustadt [31, 32], and this constraint has a variety of applications. For example, if $\|\mathbf{u}\| = \sum_{i=1}^{m} |u_i|$ and $g(\lambda) = |\lambda|$, the problem is a minimum "fuel" problem [2]. The inner product of $(\mathbf{x}(T) - \mathbf{z})$ with $\mathbf{Q}(\mathbf{x}(T) - \mathbf{z})$ is used to define the word "close."

In addition to its inherent significance, this problem formulation is of special interest, since many other optimal-control problems can be solved by successive solution of this one. The following examples will bear out this point: (1) A sequence of solutions with T^i and ξ^i, $i = 1, 2, \ldots$, converging to T^* and $\xi(\tilde{\mathbf{u}}_{(t_0,T]})$, respectively, where $\tilde{\mathbf{u}}_{(t_0,T]}$ is the time-optimal control, will solve the time-optimal problem; (2) Suppose that $T > T^*$, then a sequence of solutions with ξ^i converging to $\xi(\hat{\mathbf{u}}_{(t_0,T]})$ will solve the fixed-time minimum-effort problem (that is, where $\hat{\mathbf{u}}_{(t_0,T]}$ is the minimum-effort control) for the constraint $\mathbf{x}(T) = \mathbf{z}$; (3) The free-time version of (2) can be solved by successive solution of (2) for an increasing sequence T^j; (4) The case where the target $\mathbf{z}$ is varying with time can also be treated as in (1), (2), and (3) by successive choices of the terminal time T.

3.3 Preliminary Results

Consider the Euclidean space of (n + 1) vectors $(\mathbf{x}', \xi)'$, and let $\overline{R}_{n+1}(t_0, T, \mathbf{x}_0; \Omega)$, or simply $\overline{R}_{n+1}$, denote the set of states in R_{n+1} defined as

$$\overline{R}_{n+1} = ((\mathbf{x}', \xi)' : \mathbf{x} = \mathbf{X}(T)\mathbf{X}^{-1}(t_0)(\mathbf{x}_0 + \int_{t_0}^{T} \mathbf{X}(t_0)\mathbf{X}^{-1}(s)\mathbf{B}(s)\mathbf{u}(s)\, ds),$$

$$\xi = \int_{t_0}^{T} g(\|\mathbf{u}(s)\|)\, ds, \mathbf{u}(s) \in \Omega). \tag{3.2}$$

The set $\overline{R}_{n+1}$ is compact and convex [32] and is the set of points $(\mathbf{x}', \xi)'$ such that the point $\mathbf{x} \in \mathbf{R}_n$ can be reached at time T from $\mathbf{x}_0$ at time t_0 with control effort ξ. It follows that the set

$$\overline{R}_n = (\mathbf{x}: \mathbf{x} = \mathbf{X}(T)\mathbf{X}^{-1}(t_0)(\mathbf{x}_0 + \int_{t_0}^{T} \mathbf{X}(t_0)\mathbf{X}^{-1}(s)\mathbf{B}(s)\mathbf{u}(s)\, ds),$$

$$\mathbf{u}(s) \in \Omega, \int_{t_0}^{T} g(\|\mathbf{u}(s)\|)ds \leqslant \xi); \tag{3.3}$$

that is,

$$\bar{R}_n = \bar{R}_n(t_0, T, \mathbf{x}_0, \hat{\xi}; \Omega),$$

is a convex, compact set. Also, $\bar{R}_n$ is the set of points $\mathbf{x} \in R_n$ that can be reached from $\mathbf{x}_0$ at t_0 in the time T and with control effort $\xi \leqslant \hat{\xi}$.

Theorem 1. 2 defines a Hamiltonian for this problem for the case where the effort constraint is not present. This Hamiltonian is also significant when the effort constraint is included. We have

$$H(\mathbf{x}, \mathbf{u}, \mathbf{p}, p_0, t) = p_0 g(\|\mathbf{u}(t)\|) + \langle \mathbf{p}(t), \mathbf{A}(t)\mathbf{x}(t)\rangle + \langle \mathbf{p}(t), \mathbf{B}(t)\mathbf{u}(t)\rangle. \tag{3.4}$$

Consider the minimization of this Hamiltonian at each time t. Since the second term in Equation 3. 4 is independent of $\mathbf{u}(t)$, it is only necessary to consider the function

$$h(\mathbf{u}(t)) = p_0 g\|\mathbf{u}(t)\|) + \langle \mathbf{p}(t), \mathbf{B}(t)\ \mathbf{u}(t)\rangle. \tag{3.5}$$

Lemma 3. 1

If $\bar{\mathbf{u}}_{(t_0,T]}$ minimizes $h(\mathbf{u}(t))$ (see Equation 3. 5) almost everywhere in the interval $(t_0, T]$, then the corresponding terminal state $\bar{\mathbf{x}}(T)$ and control effort $\bar{\xi}$ are such that $(\bar{\mathbf{x}}(t)', \xi)' \in \partial\bar{R}_{n+1}$ at a point having a support hyperplane whose inward normal (that is, inward to $\bar{R}_{n+1}$) has the direction of $(\mathbf{p}'(T), p_0)'$.

Proof:

The vector $\mathbf{p}(t)$ satisfies the differential equation (Theorem 1. 2)

$$\dot{\mathbf{p}}(t) = -\mathbf{A}'(t)\mathbf{p}(t), \qquad \mathbf{p}(T) = \boldsymbol{\pi}. \tag{3.6}$$

Since $\bar{u}(t)$ minimizes $h(\mathbf{u}(t))$ (Equation 3. 5), $\bar{u}$ is a function of the time t, the final value $\boldsymbol{\pi}$, and p_0; that is, $\bar{u} = \bar{u}(t, \boldsymbol{\pi}, p_0)$. Equation 3. 1 admits the solution

$$x(t) = \mathbf{X}(t)\mathbf{X}^{-1}(t_0)(x_0 + \int_{t_0}^{T} \mathbf{X}(t_0)\mathbf{X}(s)\mathbf{B}(s)u(s)\ ds), \tag{3.7}$$

where (Appendix B)

$$\dot{X}(t) = \mathbf{A}(t)\ \mathbf{X}(t), \qquad \mathbf{X}(0) = \mathbf{I}. \tag{3.8}$$

Let $\bar{\mathbf{w}} = \bar{\mathbf{x}}(T) - \mathbf{X}(T)\mathbf{X}^{-1}(t_0)\mathbf{x}_0$, and consider the number

$$p_0\bar{\xi} + \langle\boldsymbol{\pi}, \bar{\mathbf{w}}\rangle = \int_{t_0}^{T} (\langle\boldsymbol{\pi}, \mathbf{X}(T)\mathbf{X}^{-1}(s)\mathbf{B}(s)\bar{u}(s)\rangle + p_0 g(\|\bar{u}(s)\|))ds. \tag{3.9}$$

From Equation 3. 6, we have

$$\mathbf{p}(t) = (\mathbf{X}(T)\mathbf{X}^{-1}(t))'\boldsymbol{\pi}, \tag{3.10}$$

whence we see that

$$\langle \boldsymbol{\pi}, \overline{\mathbf{w}} \rangle + p_0 \bar{\xi} = \int_{t_0}^{T} h(\bar{\mathbf{u}}(s))\, ds. \tag{3.11}$$

But, given $\boldsymbol{\pi}$ and p_0, $\bar{\mathbf{u}}(t)$ minimizes $h(\mathbf{u}(t))$, and then we have

$$\langle \pi, \overline{w} \rangle + p_0 \bar{\xi} < \int_{t_0}^{T} (\langle \boldsymbol{\pi}, \mathbf{X}(T)\mathbf{X}^{-1}(s)\mathbf{B}(s)\mathbf{u}(s) \rangle + p_0 g(\|\mathbf{u}(s)\|))ds \tag{3.12}$$

for any other admissible $\mathbf{u}_{(t_0,T]}$. Suppose that $(\bar{x}(t)', \bar{\xi})'$ is in the interior of $\overline{R}_{n+1}$, then there is an $\epsilon > 0$ (by definition of the interior) such that for all $\mathbf{x}$ and with $|x_i| \leq 1, i = 1, 2, \ldots, n$, the points $((\bar{\mathbf{x}}(T) + \epsilon\mathbf{x})', \bar{\xi} + \epsilon)'$ are also in the interior of $\overline{R}_{n+1}$. In particular, the point corresponding to $x_i = -\operatorname{sgn}(\pi_i), i = 1, 2, \ldots, n$ and $u = 0$—that is, $((\bar{\mathbf{x}}(T) + \epsilon \operatorname{sgn}(\boldsymbol{\pi}))', \bar{\xi})$—is in the interior, and hence there corresponds at least one control with this effort and which steers the state to this point. But this control must violate Relation 3.12, since

$$\langle \boldsymbol{\pi}, \overline{\mathbf{w}} \rangle - \langle \epsilon \operatorname{sgn}(\boldsymbol{\pi}), \pi \rangle + p_0 \bar{\xi} < \langle \boldsymbol{\pi}, \overline{\mathbf{w}} \rangle + p_0 \bar{\xi}. \tag{3.13}$$

We conclude that $(\bar{\mathbf{x}}(T)', \bar{\xi})' \in \partial \overline{R}_{n+1}$.

The second part of the lemma is also proved by contradiction. Suppose the contrary; that is, suppose that $(\bar{x}', \bar{\xi})' \in \partial \overline{R}_{n+1}$; $\langle \pi, \bar{\mathbf{x}}(T) \rangle + p_0 \xi <$ $\langle \boldsymbol{\pi}, \mathbf{x} \rangle + p_0 \xi$ for all $(\mathbf{x}', \xi)' \in \overline{R}_{n+1}$; but $(\pi', p_0)'$ is not normal to a support hyperplane of $\overline{R}_{n+1}$ at $(\bar{\mathbf{x}}(T)', \xi)'$. Then, the hyperplane whose normal has the direction of $(\boldsymbol{\pi}', p_0)'$ and which contains the point $(\bar{\mathbf{x}}(T)', \xi)'$ also contains points in the interior of R_{n+1}. But this implies that there are points in the interior of R_{n+1} for which $\langle \boldsymbol{\pi}, \overline{\mathbf{w}} + p_0 \bar{\boldsymbol{\xi}}$ is a constant, a situation that contradicts the first part of the lemma.

Q.E.D.

Let $\overline{P}(t_0, T, \mathbf{x}_0, \bar{\xi}; \Omega) \subseteq \overline{R}_n(t_0, T, \mathbf{x}_0, \bar{\xi}; \Omega)$, or simply $\overline{P}$, denote the set

$$\overline{P}(t_0, T, \mathbf{x}_0, \bar{\xi}; \Omega) = (\mathbf{x}{:}\ \mathbf{x} = \mathbf{X}(T)\mathbf{X}^{-1}(t_0)(\mathbf{x}_0 + \int_{t_0}^{T} \mathbf{X}(t_0)\mathbf{X}^{-1}(s)\mathbf{B}(s)\mathbf{u}(s)ds), \tag{3.14}$$

where

$$\mathbf{u}(s) \in \Omega,\ \int_{t_0}^{T} g(\|\mathbf{u}(s)\|)ds = \bar{\xi}).$$

In words, $\overline{P}$ is the projection of the intersection of $\overline{R}_{n+1}$ with the hyperplane given by the constraint $\xi = \bar{\xi}$ on the state space. It follows from Lemma 3.1 and the fact that $\overline{R}_{n+1}$ is convex, that $\overline{P}$ is convex (and compact) and $\bar{\mathbf{x}}(T)$ is on the boundary of $\overline{P}$ at the point having a support hyperplane whose inward normal is $\boldsymbol{\pi}$($\pi \neq 0$ assumed). Note that if $\boldsymbol{\pi} = 0$, the control $\mathbf{u} = \mathbf{0}$ minimizes $h(\mathbf{u}(t))$.

A few examples will serve to illustrate some of the properties of these sets.

Example 3.1. Suppose that $x(t) = u(t)$, $x(0) = 1$, $\Omega = (u: |u| \leq 1)$, $g(\|u\|) = u^2$. Then we have

$$x(T) = 1 + \int_0^T u(s)\, ds, \tag{3.15}$$

and, since

$$p(t) = \pi = \text{const}, \tag{3.16}$$

we have

$$h(u(t)) = p_0 u^2(t) + \pi u(t), \tag{3.17}$$

from which we obtain

$$\bar{u}(t) = \begin{cases} -\text{sgn}(\pi), & p_0 = 0 \quad \text{or} \quad \pi/2p_0 > 1, \\ -\pi/2p_0, & p_0 \neq 0, \quad \pi/2p_0 \leq 1. \end{cases} \tag{3.18}$$

Whence we have

$$\bar{\xi} = \begin{cases} T, & p_0 = 0 \quad \text{or} \quad \pi/2p_0 > 1, \\ \left(\dfrac{\pi}{2p_0}\right)^2 T, & p_0 \neq 0, \quad \pi/2p_0 \leq 1. \end{cases} \tag{3.19}$$

The sets $\overline{R}_{n+1}$, $\overline{R}_n$, and $\overline{P}$ are illustrated in Figure 3.1. Note that $\overline{R}_{n+1}$ is convex, compact, and symmetric about the set of points $(\mathbf{x}(0)'\xi)'$, $0 \leq \xi \leq \infty$. The symmetry follows immediately from the assumption

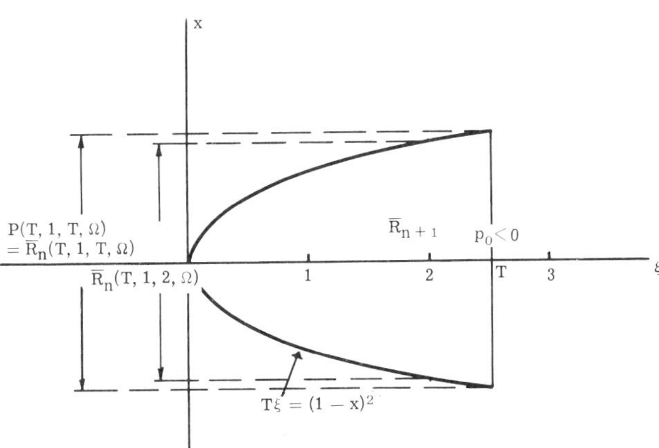

Figure 3.1 Example 3.1

of a linear plant and the fact that Ω is symmetric (see Lemma 3.2). Note also that $\overline{P}(T, \mathbf{x}_0, \xi; \Omega) = \overline{R}_n(T, \mathbf{x}_0, \xi; \Omega), 0 \leq \xi \leq T$. This is because of the fact that if $p_0 = 0, \xi$ takes on its maximum value, subject to the constraint $\mathbf{u}(t) \in \Omega$, regardless of what $\boldsymbol{\pi}$ is (see Lemma 3.3). The fact that $\overline{P} = \overline{R}_n$ for all $0 \leq \xi \leq \xi_{max}$ means that any terminal state which is reachable in time T with control effort $\xi_1 \leq \xi_2$, $0 \leq \xi_2 \leq \xi_{max}$, is also reachable in time T if the control effort is constrained to be ξ_2. In Example 3.2, these conditions are not met. Finally, note that the $\xi = T$ side of $\overline{R}_{n+1}$, excluding the two vertices which correspond to $p_0 = 0$, cannot be terminal points of extremal trajectories since $p_0 < 0$, whereas all other points on the boundary of $\overline{R}_{n+1}$ can. Furthermore, the extremal control which takes the state from $\mathbf{x}_0$ to a $(\bar{\mathbf{x}}(T)', \bar{\xi})'$ on the boundary of $\overline{R}_{n+1}$ is the optimal control for steering the state from $\mathbf{x}_0$ to that state $\mathbf{x}(T)$ with minimum control effort. In other words, if a target point $\mathbf{z}$ is in $\overline{R}_n(T, \mathbf{x}_0, \xi_{max}; \Omega)$, then there is a minimum-effort control for reaching that point. Moreover, if the extremal controls are unique, the optimal control is unique.

Example 3.2. Let the problem be as in Example 3.1, except that $\Omega = \Omega_2 = (u{:}1 \leq |u| \leq 2)$. Then, we have

$$\bar{u}(t) = \begin{cases} +2, & \pi < 0, \quad p_0 = 0; \quad \pi < -4p_0, \quad p_0 > 0, \\ +1, & \pi > 0, \quad p_0 = 0; \quad \pi > -2p_0, \quad p_0 > 0, \\ -\pi/2p_0, & 1 \leq -\pi/2p_0 \leq 2, \quad p_0 > 0. \end{cases} \tag{3.20}$$

and

$$\xi = \begin{cases} 4T, & \pi < 0, \quad p_0 = 0; \quad \pi < -4p_0, \quad p_0 > 0, \\ T, & \pi > 0, p_0 = 0; \quad \pi > -2p_0, \quad p_0 > 0, \\ (\pi/2p_0)^2 T, & 1 \leq -\pi/2p_0 \leq 2, \quad p_0 > 0. \end{cases} \tag{3.21}$$

The sets $\overline{R}_{n+1}, \overline{R}_n$, and $\overline{P}$ are shown in Figure 3.2. Note that these sets are again convex and compact, but in this case they are not symmetric about the set of points $(\mathbf{x}(0), \xi)', 0 < \xi \leq \infty$. Note also that the two points corresponding to $p_0 = 0$ require different control efforts. Finally, if $\mathbf{x}_1(T)$ is reachable with effort ξ_1, it is not necessarily reachable if the effort is constrained to be $\xi_2 \neq \xi_1$.

Lemma 3.2

If the set Ω is symmetric, then the set $\overline{R}_{n+1}(t_0, T, \mathbf{x}_0; \Omega)$ is symmetric about the set of points $(\mathbf{X}(T)\mathbf{X}^{-1}(t_0)\mathbf{x}_0, \xi)'$ in R_{n+1}.

Proof:

Suppose that $(\mathbf{x}_1(T)', \xi_1)' \in \overline{R}_{n+1}$, then there is a control with $\mathbf{u}_1(t)$ in Ω which steers the state from $\mathbf{x}_0$ to $\mathbf{x}_1(T)$ with effort ξ_1. Since the plant is linear, the control $\mathbf{u}_2(t) = -\mathbf{u}_1(t)$ steers the state to a point $\mathbf{x}_2(T)$, such that $-\mathbf{X}(T)\mathbf{X}^{-1}(t_0)\mathbf{x}_0 + \mathbf{x}_1(T) = -(\mathbf{x}_2(T) - \mathbf{X}(T)\mathbf{X}^{-1}(t_0)\mathbf{x}_0)$.

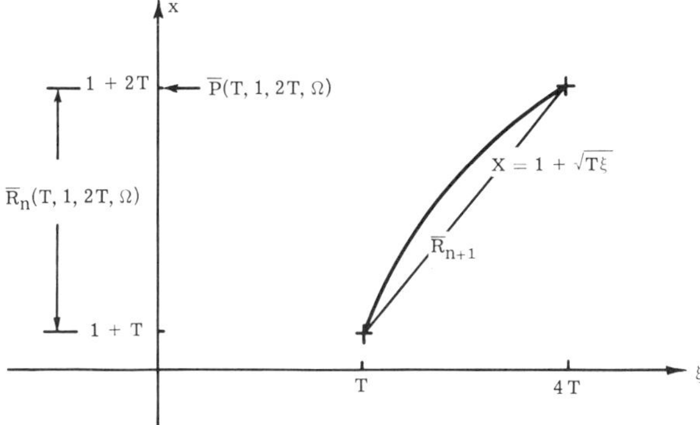

Figure 3. 2 Example 3. 2

Since Ω is symmetric, $\mathbf{u}_2(t)$ is admissible, and since $\|-\mathbf{u}\| = \|\mathbf{u}\|$, $\xi_1 = \xi_2$. It follows that $(\mathbf{X}(T)\mathbf{X}^{-1}(t_0)\mathbf{x}_0', \xi)', -\infty < \xi < \infty$ is an axis of symmetry. Q.E.D.

Lemma 3. 3

If the value of ξ corresponding to $p_0 = 0$, call it ξ_0, is maximal—that is, if $\xi_0 = \max_{\mathbf{u}(t) \in \Omega} \int_{t_0}^{T} g(\|\mathbf{u}(s)\|)ds$—and is independent of the initial condition $\boldsymbol{\pi}$ on $\mathbf{p}(t)$ (see Equation 3. 6), then $\overline{P}(t_0, T, \mathbf{x}_0, \xi; (\Omega) = \overline{R}_n(t_0, T, \mathbf{x}_0, \xi; \Omega)$.

Proof:

The set $\overline{R}_{n+1}$ is convex and compact, and if the conditions of the lemma hold; then $\overline{R}_{n+1}$ has a portion of its boundary contained in the hyperplane defined by $\xi = \xi_0$. This portion contains all the points on the boundary of $\overline{R}_{n+1}$ for which one can have $p_0 = 0$. It follows that $p_0 > 0$ for every point on $\partial\overline{R}_{n+1}$ with $\xi < \xi_0$. But if this is so, $\overline{P} = \overline{R}_n$. (See Figure 3. 1, for example.)

In other words, if the conditions of the lemma are satisfied and $\mathbf{x}$ is reachable with effort ξ_1, then it is also reachable if the effort is constrained to be $\xi_2 > \xi_1$. Most of the more popular constraint sets Ω appearing in the literature are such that Lemma 3. 3 holds. For example, if $\Omega = (\mathbf{u}: \|\mathbf{u}\| \leq 1)$, where $\|\mathbf{u}\|$ means any norm, then regardless of $\boldsymbol{\pi}$, $\xi_0 = (T - t_0)g(1)$ if $\mathbf{p}_0 = 0$. Q.E.D.

Lemma 3. 4

Let (1) $\mathbf{c}(t) = \mathbf{B}'(t)\mathbf{p}(t)$, $\mathbf{p}(t)$ admissible, $t \in (t_0, T]$, and (2) $\bar{\xi}(\boldsymbol{\pi}, \alpha, p_0)$, $\boldsymbol{\pi} = \mathbf{p}(T)$ denote the effort expended by the control $\bar{\mathbf{u}}_{(t_0, T]}$ that minimizes $p_0 g(\|\mathbf{u}(t)\|) + \alpha \langle \mathbf{c}(t), \mathbf{u}(t)\rangle$ uniquely almost everywhere in $(t_0, T]$. Then, we see that (1) $\bar{\xi}(\boldsymbol{\pi}, \alpha, 1)$ is a nondecreasing function of α, and (2) $\lim_{\alpha \to \infty} \bar{\xi}(\boldsymbol{\pi}, \alpha, 1) = \bar{\xi}(\boldsymbol{\pi}, \alpha, 0)$, which is independent of $\alpha > 0$.

Proof:

Let $\bar{u}_i$ minimize $p_0 g(\|\mathbf{u}(t)\|) + \alpha_i \langle \mathbf{c}(t), \mathbf{u}(t) \rangle$, $i = 1, 2$, $\alpha_1 > \alpha_2$. Then, we have

$$p_0 g(\|\bar{\mathbf{u}}_1(t)\|) + \alpha_1 \langle \mathbf{c}(t), \bar{\mathbf{u}}_1(t) \rangle \leq p_0 g(\|\bar{\mathbf{u}}_2(t)\|) + \alpha_1 \langle \mathbf{c}(t), \bar{\mathbf{u}}_2(t) \rangle, \tag{3. 22}$$

and

$$p_0 g(\|\bar{\mathbf{u}}_2(t)\|) + \alpha_2 \langle \mathbf{c}(t), \bar{\mathbf{u}}_2(t) \rangle \leq p_0 g(\|\bar{\mathbf{u}}_1(t)\|) + \alpha_2 \langle \mathbf{c}(t), \bar{\mathbf{u}}_1(t) \rangle. \tag{3. 23}$$

Multiply Equation 3. 22 by α_2 and Equation 3. 23 by α_1 and add to obtain

$$p_0(\alpha_2 - \alpha_1)\,(g(\|\bar{\mathbf{u}}_1(t)\|) - g(\|\bar{\mathbf{u}}_2(t)\|)) \leq 0. \tag{3. 24}$$

From Equation 3. 24, we conclude that

$$g(\|\bar{\mathbf{u}}_1(t)\|) \geq g(\|\bar{\mathbf{u}}_2(t)\|), \tag{3. 25}$$

unless $p_0 = 0$.

Suppose that $p_0 = 0$. Then, it is easily seen that if $\alpha > 0$, $\bar{\mathbf{u}}_i(t)$ lies on the boundary Ω at a point such that the support hyperplane to Ω at that point has a normal, inward to Ω, whose direction is that of $\mathbf{c}(t)$. If this control is unique almost everywhere, then $\bar{\xi}(\pi, \alpha, 0)$ is independent of α and clearly is the same as $\lim_{\alpha \to \infty} \bar{\xi}(\pi, \alpha, 1)$, since letting $\alpha \to \infty$ has the same effect on $\bar{u}(t)$ as letting $p_0 = 0$. Q.E.D.

Setting $p_0 = 0$ yields controls which are similar to time-optimal controls. Indeed, if $p_0 = 0$, the set of controls generated by selecting π's and minimizing $h(u(t))$ will include the set of time-optimal controls, in the sense that the set of terminal states which can be reached in a time-optimal fashion with terminal time T will be included in the set of states for which one can have $p_0 = 0$. However, not all of the controls generated by setting $p_0 = 0$, etc., will necessarily be time-optimal controls. For example, in Figure 3. 1 the controls that steer the state to either of the two points at which it is possible to have $p_0 = 0$ are both time-optimal. However, only one of the corresponding controls in Figure 3. 2 is a time-optimal control.

Lemma 3. 5

A. If (1) $\bar{\mathbf{u}}_1(t)$ minimizes $g(\|\mathbf{u}(t)\|) + \alpha \langle \mathbf{c}(t), \mathbf{u}(t) \rangle$, $\alpha > 0$, and $\mathbf{c}(t)$ admissible, for each t in $(t_0, T]$, and (2) the effort $\xi(\bar{u}_{1(t_0,T]})$ expended by the control $\bar{\mathbf{u}}_{1(t_0,T]}$ and that expended by $\mathbf{u}_{2(t_0,T]}$, $\mathbf{u}_2(t) \in \Omega$ satisfy the relation

$$\xi(\bar{\mathbf{u}}_{1(t_0,T]}) \geqslant \xi(\mathbf{u}_{2(t_0,T]}), \tag{3.26}$$

then we have

$$\int_{t_0}^{T} \langle \mathbf{c}(s), (\bar{\mathbf{u}}_1(s) - \mathbf{u}_2(s))\rangle \, ds \leqslant 0. \tag{3.27}$$

B. If $\hat{\mathbf{u}}_1$ minimizes $\langle \mathbf{c}(t), \mathbf{u}(t)\rangle$ for each t in $(t_0, T]$ and $\mathbf{u}_2(t)$ is admissible, then Equation 3. 27 holds with $\bar{\mathbf{u}}_1(s)$ replaced by $\hat{\mathbf{u}}_1(s)$. Further, if the problem is normal, the equality holds in Equation 3. 27 if and only if $\bar{\mathbf{u}}_1(s) = \mathbf{u}_2(s)$ or $\hat{\mathbf{u}}_1(s) = \mathbf{u}_2(s)$ (whichever is applicable) almost everywhere in the interval $(t_0, T]$.

Proof:

From Assumption (1), we see that

$$p_0 g(\|\bar{\mathbf{u}}_1(t)\|) + \alpha\langle \mathbf{c}(t), \bar{\mathbf{u}}_1(t)\rangle \leqslant p_0 g(\|\mathbf{u}_2(t)\|) + \alpha\langle \mathbf{c}(t), \mathbf{u}_2(t)\rangle. \tag{3.28}$$

Hence we have

$$\int_{t_0}^{T} \langle \mathbf{c}(s), \bar{\mathbf{u}}_1(s) - \mathbf{u}_2(s)\rangle \, ds \leqslant (\xi(\mathbf{u}_{2(t_0,T]}) - \xi(\bar{\mathbf{u}}_{1(t_0,T]}))p_0. \tag{3.29}$$

But, $p_0 = 1$ for Case A, and $p_0 = 0$ for Case B. Hence, Equation 3. 27 follows in any case. If the problem is normal, $\bar{\mathbf{u}}_1(t)$ is unique except at perhaps a finite number of times in the interval $(t_0, T]$, and hence Equation 3. 28 holds with the strict inequality almost everywhere. It follows immediately that Equation 3. 29 holds with a strict inequality, and hence Equation 3. 27 does too, unless, of course, $\bar{\mathbf{u}}_1(s) = \mathbf{u}_2(s)$ or $\hat{\mathbf{u}}_1(s) = \mathbf{u}_2(s)$. Q.E.D.

Lemma 3. 6

If the problem is normal and $\mathbf{x}(T) \in \overline{P}(t_0, T, \mathbf{x}_0, \xi; \Omega)$ at a point such that $p_0 \geqslant 0$, the control which steers the state from $\mathbf{x}_0$ at t_0 to $\mathbf{x}(T)$ is unique (that is, only controls which minimize the Hamiltonian steer the state to points on the boundary of P where $p_0 \geqslant 0$, and these controls are unique.)

Proof:

Suppose that there are two controls, $\bar{\mathbf{u}}_{1(t_0,T]}$ and $\mathbf{u}_{2(t_0,T]}$, both of which steer the state to $\mathbf{x}(T)$. It is already known (Lemma 3. 1) that there is a control which minimizes $p_0 g(\|\mathbf{u}(t)\|) + \langle \mathbf{B}'(t)\mathbf{p}(t), \mathbf{u}(t)\rangle$, $\mathbf{p}(t)$ admissible, and which steers the state to $\mathbf{x}(T)$. Further, $\mathbf{p}(T)$ is the inward normal to a support hyperplane of $\overline{P}(t_0, T, \mathbf{x}_0, \xi; \Omega)$ at $\mathbf{x}(T)$. Let $\bar{\mathbf{u}}_{1(t_0,T]}$ be this control. Then, since both controls steer the state to $\mathbf{x}(T)$, we have

$$\int_{t_0}^{T} \mathbf{X}(T)\mathbf{X}^{-1}(s)\mathbf{B}(s)(\bar{\mathbf{u}}_1(s) - \mathbf{u}_2(s))ds = \mathbf{0}. \tag{3.30}$$

Hence, we have

$$\int_{t_0}^{T} < \pi, \mathbf{X}(T)\mathbf{X}^{-1}(s)\mathbf{B}(s)(\bar{\mathbf{u}}_1(s) - \mathbf{u}_2(s)) > ds = 0, \tag{3.31}$$

which contradicts Lemma 3.5, since $\mathbf{p}(t) = (\mathbf{X}(T)\mathbf{X}^{-1}(t))'\pi$. Q.E.D.

In other words, only the controls that minimize $h(\mathbf{u}(t))$ and expend effort ξ cause the state to reach the boundary of $\overline{P}$ at points where $p_0 \geq 0$, and these controls are unique. This means that even though $\overline{P}$ might have many support hyperplanes at a given point, all of the costate vectors corresponding to these hyperplanes will yield the same control via the minimization of $h(\mathbf{u}(t))$.

The question of normality has been brought up several times. We treat this matter in some detail in Appendix C.

Lemma 3.7

Let $\mathbf{c}(t)$ be admissible, and let $\bar{\mathbf{u}}(t)$ and $\hat{\mathbf{u}}(t)$ be the vectors in Ω which minimize $\langle \mathbf{c}(t), \mathbf{u}(t)\rangle + g(\|\mathbf{u}(t)\|)$ and $\langle \mathbf{c}(t), \mathbf{u}(t)\rangle$, respectively. Then, $\xi(\bar{\mathbf{u}}_{(t_0,T]}) \leq \xi(\hat{\mathbf{u}}_{(t_0,T]})$.

Proof:

Since $\bar{\mathbf{u}}(t)$ minimizes $g(\|\mathbf{u}(t)\|) + \langle \mathbf{c}(t), \mathbf{u}(t)\rangle$, we find that

$$g(\|\bar{\mathbf{u}}(t)\|) + \langle \mathbf{c}(t), \bar{\mathbf{u}}(t)\rangle \leq g(\|\hat{\mathbf{u}}(t)\|) + \langle \mathbf{c}(t), \hat{\mathbf{u}}(t)\rangle,$$

whence we have

$$g(\|\bar{\mathbf{u}}(t)\|) - g(\|\hat{\mathbf{u}}(t)\|) \leq \langle \mathbf{c}(t), \hat{\mathbf{u}}(t) - \bar{\mathbf{u}}(t)\rangle.$$

But, $\hat{\mathbf{u}}(t)$ minimizes $\langle \mathbf{c}(t), \mathbf{u}(t)\rangle$, which implies that $\langle \mathbf{c}(t), \hat{\mathbf{u}}(t) - \bar{\mathbf{u}}(t)\rangle \leq 0$. Therefore, we see that $g(\|\bar{\mathbf{u}}(t)\|) \leq g(\|\hat{\mathbf{u}}(t)\|)$.

Q.E.D.

3.4 The Iterative Procedure

Let T^* denote the least terminal time for the problem of steering the state from $\mathbf{x}_0$ to $\mathbf{z}$ with $\mathbf{u}(t)$ in Ω (that is, assuming the existence of a time-optimal control) and assuming that $T \leq T^*$. Let $\tilde{\mathbf{u}}_{(t_0,T]}$ denote a control which minimizes $\langle \mathbf{x}(T) - \mathbf{z}, \mathbf{Q}(\mathbf{x}(T) - \mathbf{z})\rangle$ when there is no effort constraint; that is, $\tilde{\mathbf{u}}_{(t_0,T]}$ is the solution to Problem Z_5^1 (see Section 2.3) which is also assumed to exist. Finally, assuming that $T > T^*$, let $\tilde{\mathbf{u}}_{(t_0,T]}$ denote a control which steers the state from $\mathbf{x}_0$ to $\mathbf{z}$ and minimizes the effort ξ. Then, it is assumed that T and $\hat{\xi}$ (see Section 3.2) satisfy the following conditions:

$$\text{If } T \leqslant T^*, \qquad \hat{\xi} \leqslant \xi(\bar{\mathbf{u}}_{(t_0,T]}). \tag{3.32}$$

$$\text{If } T > T^*, \qquad \hat{\xi} \leqslant \xi(\tilde{\bar{\mathbf{u}}}_{(t_0,T]}). \tag{3.33}$$

The procedure is stated for the most general problem first, and it is then simplified for the more special case (that appearing with some regularity in the literature) satisfying Lemma 3.3.

Now we let (1) the superscript k denote the kth interation; (2) $\mathbf{x}^k(T)$ be the terminal state after the application of the control $\mathbf{u}^k_{(t_0,T]}$; (3) $\mathbf{c}^k(t) = \mathbf{B}'(t)(\mathbf{X}(T)\mathbf{X}^{-1}(t))'\mathbf{Q}\mathbf{y}^k$, where $\mathbf{y}^k = \mathbf{x}^k(T) - \mathbf{z}$; (4) $\hat{\mathbf{u}}^k_{(t_0,T]}$ be the control that minimizes $\langle \mathbf{c}^k(t), \mathbf{u}(t)\rangle$ (that is, $p_0 = 0$); (5) $\bar{\mathbf{u}}^k_{(t_0,T]}$ be the control that minimizes $g(\|\mathbf{u}(t)\|) + \alpha^k\langle \mathbf{c}^k(t), \mathbf{u}(t)\rangle$ with α^k chosen so that $\xi(\bar{\mathbf{u}}^k_{(t_0,T]}) = \hat{\xi}$; (6) $\bar{\mathbf{x}}^k(T)$ and $\hat{\mathbf{x}}^k(T)$ be the terminal states after the application of the controls $\bar{\mathbf{u}}^k_{(t_0,T]}$ and $\hat{\mathbf{u}}^k_{(t_0,T]}$, respectively; (7) $\mathbf{w}^k = \hat{\mathbf{x}}^k(T) - \mathbf{x}^k(T)$ if $\xi(\hat{\mathbf{u}}^k_{(t_0,T]}) = \hat{\xi}$ and $\mathbf{w}^k = \bar{\mathbf{x}}^k(T) - \mathbf{x}^k(T)$ otherwise; and finally (8) let $\lambda^k = \text{sat}(\langle -\mathbf{y}^k, \mathbf{Q}\mathbf{w}^k\rangle / \langle \mathbf{w}^k, \mathbf{Q}\mathbf{w}^k\rangle)$, where $\text{sat}(x) = x$ if $|x| \leqslant 1$ and $= 1$ if $|x| > 1$.

The procedure is initialized using $\mathbf{u}^0(t) = \mathbf{d}, \mathbf{d} \in \Omega$, a constant, and where $\mathbf{d}$ is chosen so that $\xi(\mathbf{u}^0_{(t_0,T]}) = g(\|\mathbf{d}\|)(T - t_0) \leqslant \hat{\xi}$. ($\mathbf{d} = \mathbf{0}$ is the recommended easy choice when $\mathbf{0} \in \Omega$). The following steps then describe the iterative procedure:

Step 1: Find $\mathbf{x}^k(T)$ and $\mathbf{y}^k$.

Step 2: Find $\hat{\mathbf{u}}^k_{(t_0,T]}$ and if $\xi(\hat{\mathbf{u}}^k_{(t_0,T]}) \leqslant \hat{\xi}$, use $\mathbf{u}^{k+1}(t) = (1 - \lambda^k)\mathbf{u}^k(t) + \lambda^k\hat{\mathbf{u}}^k(t)$, provided that $\langle \mathbf{y}^k, \mathbf{Q}\mathbf{w}^k\rangle \neq 0$. If $\xi(\hat{\mathbf{u}}^k_{(t_0,T]}) > \hat{\xi}$, proceed to Step 3.

Step 3: Find $\bar{\mathbf{u}}^k_{(t_0,T]}$ and $\bar{\mathbf{x}}^k(T)$, and use

$$\mathbf{u}^{k+1}(t) = (1 - \lambda^k)\,\mathbf{u}^k(t) + \lambda^k\bar{\mathbf{u}}^k(t), \tag{3.34}$$

and return to Step 1 unless $\langle \mathbf{y}^k, \mathbf{Q}\mathbf{w}^k\rangle = 0$, in which case the iteration ends.

It is assumed that the problem is normal (see Appendix C), and hence it is always possible to determine $\hat{\mathbf{u}}^k(t)$. Also, it follows from Lemma 3.4 and the normality assumption that $\bar{\mathbf{u}}^k(t)$ can always be found when $\xi(\hat{\mathbf{u}}^k_{(t_0,T]}) > \hat{\xi}$. In other words, all of the steps can be carried out.

From a computational point of view, Step 3 will be the most difficult, and the degree of difficulty experienced in carrying out both Steps 3 and 4 will depend heavily on the set Ω.

3.5 Convergence

Theorem 3.1

If the problem is normal, the iterative procedure has the property that the sequence $\mathbf{u}^k(t_0, T]$, $k = 0, 1, 2, \ldots,$ converges to $\mathbf{u}^*(t_0, T]$, the control that minimizes $\langle \mathbf{y}, \mathbf{Q}\mathbf{y} \rangle$ with $\xi(\mathbf{u}^*(t_0, T]) \leqslant \hat{\xi}$. Moreover, the sequence $\langle \mathbf{y}^k, \mathbf{Q}\mathbf{y}^k \rangle$ is monotone decreasing.

Proof:

Since either

$$\mathbf{u}^{k+1}(t) = (1 - \lambda^k)\,\mathbf{u}^k(t) + \lambda^k \bar{\mathbf{u}}^k(t) \tag{3.35}$$

or

$$\mathbf{u}^{k+1}(t) = (1 - \lambda^k)\mathbf{u}^k(t) + \lambda^k \hat{\mathbf{u}}^k(t), \tag{3.36}$$

and

$$\mathbf{y}^{k+1} = \mathbf{y}^k + \lambda^k \mathbf{w}^k, \tag{3.37}$$

and since the plant is linear, we have

$$\langle \mathbf{y}^{k+1}, \mathbf{Q}\mathbf{y}^{k+1} \rangle = \langle \mathbf{y}^k, \mathbf{Q}\mathbf{y}^k \rangle + 2\lambda^k \langle \mathbf{y}^k, \mathbf{Q}\mathbf{w}^k \rangle + (\lambda^k)^2 \langle \mathbf{w}^k, \mathbf{Q}\mathbf{w}^k \rangle . \tag{3.38}$$

Hence, we see that

$$\langle \mathbf{y}^{k+1}, \mathbf{Q}\mathbf{y}^{k+1} \rangle - \langle \mathbf{y}^k, \mathbf{Q}\mathbf{y}^k \rangle < 0, \tag{3.39}$$

and Relation 3.39 is minimized for $0 < \lambda^k \leqslant 1$ if

$$\lambda^k = \mathrm{sat}(-\langle \mathbf{y}^k, \mathbf{Q}\mathbf{w}^k \rangle \,/\, \langle \mathbf{w}^k, \mathbf{Q}\mathbf{w}^k \rangle), \tag{3.40}$$

provided that $\langle \mathbf{y}^k, \mathbf{Q}\mathbf{w}^k \rangle \leqslant 0$.
It follows from the definition of $\mathbf{w}^k$ that either

$$\langle \mathbf{y}^k \mathbf{Q}\mathbf{w}^k \rangle = \int_{t_0}^{T} \langle \mathbf{c}^k(t), \bar{\mathbf{u}}^k(t) - \mathbf{u}^k(t) \rangle \, dt \tag{3.41}$$

or

$$\langle \mathbf{y}^k, \mathbf{Q}\mathbf{w}^k \rangle = \int_{t_0}^{T} \langle \mathbf{c}^k(t), \hat{\mathbf{u}}^k(t) - \mathbf{u}^k(t) \rangle \, dt. \tag{3.42}$$

If $\xi(\hat{\mathbf{u}}(t_0, T]) \leqslant \hat{\xi}$, $\langle \mathbf{y}^k, \mathbf{Q}\mathbf{w}^k \rangle < 0$, unless $\mathbf{u}^k(t) = \hat{\mathbf{u}}^k(t)$ almost everywhere (see Lemma 3.5). Suppose that $\xi(\hat{\mathbf{u}}(t_0, T]^k) > \hat{\xi}$, then again we will have $\langle \mathbf{y}^k, \mathbf{Q}\mathbf{y}^k \rangle < 0$, provided that $\xi(\mathbf{u}(t_0, T]^k) \leqslant \xi(\bar{\mathbf{u}}(t_0, T]^k) = \hat{\xi}$ (Lemma 3.5).

But $\xi\,(\mathbf{u}_{(t_0,T]}{}^0) \leqslant \hat{\xi}$ and $\mathbf{u}^1(t) = (1-\lambda^0)\mathbf{u}^0(t) + \lambda^0\,\bar{\hat{\mathbf{u}}}^0(t)$, where $\bar{\hat{\mathbf{u}}}(t)$ means $\hat{\mathbf{u}}(t)$ if $\xi\,(\hat{\mathbf{u}}_{(t_0,T]}) \leqslant \hat{\xi}$ and $\bar{\mathbf{u}}(t)$ if $\xi\,(\hat{\mathbf{u}}_{(t_0,T]})$ is otherwise. Then, using the triangle inequality for norms and the monotonicity of $g(\cdot)$, we find that

$$\xi(\mathbf{u}_{(t_0,T]}{}^1) \leqslant (1-\lambda^0)\;\xi\;(\mathbf{u}_{(t_0,T]}{}^0) + \lambda^0\;\xi(\mathbf{u}_{(t_0,T]}) \leqslant \xi.$$

It follows by induction on k that

$$\xi\,(\mathbf{u}_{(t_0,T]}{}^k) \leqslant \xi\,(\bar{\mathbf{u}}_{(t_0,T]}{}^k) = \hat{\xi}.$$

Therefore, $\langle \mathbf{y}^k, \mathbf{Q}\mathbf{w}^k\rangle < 0$, unless $\mathbf{u}^k(t) = \bar{\mathbf{u}}^k(t)$ almost everywhere when $\xi\,(\hat{\mathbf{u}}_{(t_0,T]}{}^k) > \xi$. Hence, the sequence $\langle \mathbf{y}^k, \mathbf{Q}\mathbf{y}^k\rangle$ decreases monotonically until $\mathbf{u}^k(t)$ coincides with a control which steers the state to $\partial\bar{\mathbf{R}}_n(t_0, T, \mathbf{x}_0, \hat{\xi}; \Omega)$ and minimizes $h(\mathbf{u}(t))$. Furthermore, this particular terminal state is at a point on $\partial\bar{R}_n$ whose support hyperplane has the inward normal $\mathbf{Q}(\mathbf{x}^k(T) - z)$. It follows immediately that $\mathbf{x}^k(T)$ is the point on $\bar{R}_n$ that minimizes $\langle \mathbf{y}, \mathbf{Q}\mathbf{y}\rangle$ on $\partial\bar{\mathbf{R}}_n$ and hence the theorem is proved.

The iterative procedure is illustrated in Figure 2.1, where a hypothetical set $\bar{R}_n$ is illustrated. The point $x^0(T)$ is in the interior of $\bar{R}_n$. The vector $\mathbf{y}^0$ determines a point $\bar{x}^0(T)$ on $\partial\bar{R}_n$ that yields $\mathbf{w}^0$. The choice of $\lambda^1 = \mathrm{sat}(-\langle \mathbf{y}^0, \mathbf{Q}\mathbf{w}^0\rangle \,/\, \langle \mathbf{w}^0, \mathbf{Q}\mathbf{w}^0\rangle)$ ($\mathbf{Q} = \mathbf{I}$ in Figure 2.1) ensures that $\mathbf{u}^1(t)$ is admissible (since Ω is convex) and gives the point $\mathbf{x}^1(T)$, which is the point on the line segment joining $\mathbf{x}^0(T)$ and $\bar{x}^0(T)$ which is closest to $\mathbf{z}$.

Theorem 3.1 treats only the convergence. Gilbert has shown in an excellent paper [17][1] that the rate of convergence becomes, after a sufficient number of iterations, proportional to $(k)^{-1/2}$.

Suppose that the conditions of Equations 3.32 and 3.33 are not met. In other words, suppose that the target point $\mathbf{z}$ is in the interior of $\bar{R}_n(t_0, T, \mathbf{x}_0, \xi; \Omega)$. The question then is: Will the iterative procedure converge, and if so, to what? It was seen in Section 2.3 that for problem Z_5^1 under these conditions the procedure converged to a feasible control (that is, a control which steers the state to the target in the allotted time, etc., but which is not necessarily optimal). Actually, if the same situation exists here, the corresponding control will be a solution to Problem Z_5 as stated, since it is required only that $\xi \leqslant \hat{\xi}$, and this will be the case. On the other hand, since the control will not be an extremal one in the sense of the Minimum Principle, one cannot use a sequence of solutions of this type to produce solutions to other optimal-control problems as suggested in Section 3.2, when Equations 3.32 and 3.33 are not satisfied.

It is easily seen that the situation here is indeed the same as that in

---

[1] The work presented here was done independently of Gilbert's work in Reference 17.

Section 2.3. That is, Lemma 3.5 does not depend on the assumptions of Equations 3.32 and 3.33. One concludes that when these conditions are not met, the iterative procedure will converge to a feasible control.

A Fortran II program is presented in Appendix I for solving these problems when (1) **A** and **B** are constant matrices, (2) $\|\mathbf{u}\| = \sum_{i=1}^{m} |u_i|$, (3) $g(x) = |\mathbf{x}|$, and (4) $\Omega = (\mathbf{u}: |u_i| \leq 1, i = 1, 2, \ldots, m)$. This program was written for use on an IBM 1620 digital computer, and experiments were run to test it in the Computing Center at the Royal Military College of Canada. Some of these results are reported in Chapter 8.

If the reader refers to Section 2.3, he will find that this iterative procedure reduces to the one presented there if one removes Step 3.

There are also situations in which one can bypass Step 2. For example, if Lemma 3.3 is satisfied (it is in the case programmed in Appendix I) and $\hat{\xi} < \xi_0$, it is not necessary to carry out Step 2.

3.6 The Unconstrained Control

Suppose that $\Omega = R_m$; in other words, suppose that it is practical to assume that there is no limitation on $\mathbf{u}(t)$ except $\|\mathbf{u}(t)\| < \infty$. Let $N(\rho) = (\mathbf{u}: \|\mathbf{u}\| \leq \rho)$. It is shown in Appendix C that $g(\|\mathbf{u}(t)\|) + \langle \mathbf{p}(t), \mathbf{u}(t) \rangle$, $\mathbf{p}(t)$ admissible, attains a unique minimum in R_m if $g(\cdot)$ is strictly convex down and $\mathbf{p}(t)$ is not perpendicular to a line segment of $\partial N(1)$. It follows that the iterative procedure presented here can be used when $g(\cdot)$ is strictly convex down and $N(1)$ is strictly convex, or if it can be shown that $\mathbf{p}(t)$ is normal to a line segment of $\partial N(1)$ at most a finite number of times in the interval $(t_0, T]$.

3.7 Summary

Ho [19] has proposed and iterative procedure for computing the control that minimizes $\langle \mathbf{x}(T), \mathbf{x}(T) \rangle$ for a linear plant with a single input under the constraint $|u(t)| \leq 1$. Fancher [15] has improved this procedure for the same problem (see Section 2.3). The approach was extended in this chapter to the more general problem, where the problem is to minimize $\langle (\mathbf{x}(T) - \mathbf{z}), \mathbf{Q}(\mathbf{x}(T) - \mathbf{z}) \rangle$, $\mathbf{z} \in R_n$, for a linear plant having any finite number of inputs under the constraints $\mathbf{u}(t) \in \Omega$ (the origin not necessarily interior to Ω), Ω convex and compact, and $\xi(\mathbf{u}_{(t_0,T]}) \leq \hat{\xi}$. The improvement made by Fancher is preserved in this extension. It was argued that successive solutions of this problem can be used to solve other optimal-control problems. It was shown in Section 3.5 that the procedure converges to the optimal control with an unknown rate.[2] It was also shown that conditions exist under which the procedure can be used when the control is unconstrained (Section 3.6).

---

[2] See Reference 17 for a discussion on the rate of convergence and other details.

4. The Time-Optimal Problem (Ω_a)

4.1 Introduction

The time-optimal control problem Z_1 was defined in Section 1.5. In this chapter this problem is studied in some detail. The function $\mathbf{M}(v, T)$ is defined in Section 4.2 in the same way as it was in Section 2.8, where $\mathbf{M}$ was defined for a general problem. In Section 4.3, the set D of elements $(\mathbf{v}, T)$ in the domain of $\mathbf{M}(\mathbf{v}, T)$, which correspond to optimal trajectories, is defined. It is shown that $\mathbf{M}(\mathbf{v}, T)$ is one-to-one, continuous, and piecewise differentiable to all orders on D. Special attention is given to a set Q upon which the derivatives do not exist. The iterative procedure is stated formally in Section 4.4. Section 4.5 contains an investigation into the convergence of the iterative procedure, and conditions guaranteeing monotone or almost monotone convergence are given. An initial-guess procedure is proposed in Section 4.5, and in Section 4.7 the results of this chapter are generalized to the multi input case. Section 4.8 contains a discussion of the time-varying case. The chapter is concluded with a discussion, by way of an example, of singular time-optimal-control problems in Section 4.9 and a comparison in Section 4.10 of the modified and unmodified time-optimal-regulator problems with regard to the iterative procedure of Section 4.4.

4.2 The Function M

Since the input u is bounded in magnitude, the state of the plant cannot change instantaneously. Consequently, since the time is the cost, the terminal state $\mathbf{x}^*(T^*)$ of a time-optimal trajectory from any initial state in C/S (see Definition 1.6) is on the boundary ∂S of S. Then, the transversality condition NC10 of Theorem 1.2 applies to problem Z_1; that is, we have

$$\mathbf{x}^*(T^*) = v^*, \qquad v^* \in \partial S, \qquad v^* \neq 0, \tag{4.1}$$

and

$$\mathbf{p}^*(\mathrm{T}^*) = \alpha \mathbf{Q} \mathbf{v}^*, \qquad \alpha > 0. \tag{4.2}$$

From Definition 1.11 and using Equation 1.11, we find that the Hamiltonian for Problem Z_1 is

$$\mathrm{H} = p_0 + \langle \mathbf{p}(t), \mathbf{A}\mathbf{x}(t) \rangle + \langle \mathbf{b}, \mathbf{p}(t) \rangle\, u,^1 \tag{4.3}$$

whence we have

$$\dot{\mathbf{p}}^*(t) = -\mathbf{A}' \mathbf{p}^*(t), \qquad \mathbf{p}^*(\mathrm{T}^*) = \alpha \mathbf{Q} \mathbf{v}^*, \tag{4.4}$$

$$\dot{\mathbf{x}}^*(t) = \mathbf{A}\mathbf{x}^*(t) + \mathbf{b}u^*(t), \qquad \mathbf{x}^*(0) = \mathbf{x}. \tag{4.5}$$

Since $u \in \Omega_a$ (that is, $|u| \leq 1$), from NC7, we have

$$u^*(t) = -\mathrm{sgn}\{\langle \mathbf{p}^*(t), \mathbf{b} \rangle\}. \tag{4.6}$$

Given any $\mathbf{p}(\mathrm{T}) = \alpha \mathbf{Q}\mathbf{v}$, $\mathbf{v} \in R_n$, Equation 4.4 has the unique solution

$$\mathbf{p}(t) = \alpha\, e^{-\mathbf{A}'(t-\mathrm{T})} \mathbf{Q}\mathbf{v}. \tag{4.7}$$

Hence, we have

$$u^*(t) = -\mathrm{sgn}\{\alpha \langle \mathbf{v}^*, \mathbf{Q} e^{\mathbf{A}(\mathrm{T}-t)} \mathbf{b} \rangle\}. \tag{4.8}$$

Since the problem is assumed normal (see Theorem C.1, Appendix C for conditions), the control $u^*_{(0,\mathrm{T}]}$ is well defined except at a finite number of switch times. Equations 4.4, 4.5, and 4.8 are independent of $|\alpha|$ and p_0, and hence these parameters are ignored in defining $\mathbf{M}$. To define $\mathbf{M}$, choose $\mathbf{v} \neq \mathbf{0}$ in R_n and $\mathrm{T} \in R_1^+$, and let $\tau = \mathrm{T} - t$, then the initial-value problem defining $\mathbf{M}$ is

$$\mathbf{x}(\tau) = -\mathbf{A}\mathbf{x}(\tau) - \mathbf{b}u(\tau), \qquad \mathbf{x}(\tau = 0) = \mathbf{v}, \tag{4.9}$$

$$u(\tau) = -\mathrm{sgn}(\langle \mathbf{v}, \mathbf{Q} e^{\mathbf{A}\tau} \mathbf{b} \rangle). \tag{4.10}$$

Definition 4.1 (M)

Let $\mathbf{M}$ be a function whose domain is $S \times R_1^+$ and which is defined as

$$\mathbf{M}(\mathbf{v}, \mathrm{T}) = e^{-\mathbf{A}\mathrm{T}}(\mathbf{v} + \int_0^{\mathrm{T}} e^{\mathbf{A}s} \mathbf{b}\, \mathrm{sgn}(\langle \mathbf{v}, \mathbf{Q} e^{\mathbf{A}s} \mathbf{b} \rangle)\, \mathbf{ds}. \tag{4.11}$$

In other words, $\mathbf{M}(\mathbf{v}, \mathrm{T})$ is the solution of Equations 4.9 and 4.10 at $\tau = \mathrm{T}$.

---

[1] The single-input case is treated first; the results for the vector-input case are given in Section 4.7.

4.3 Definitions and Preliminary Theorems

Since the domain of **M** is restricted to $\partial S \times R_1^+$, **M** is a mapping from the n-space $\partial S \times R_1^+$ into the n-space R_n/S. It is desirable that the range and domain of **M** have the same dimension, since otherwise **M** will not be one-to-one.

Loosely speaking, an iterative procedure for inverting **M** will involve choosing elements $(\mathbf{v}, T)$ in the domain and checking to see if the particular choice satisfies $\mathbf{x}_0 = \mathbf{M}(\mathbf{v}, T)$. If **M** is one-to-one on $\partial S \times R_1^+$ and onto C/S, and if the time-optimal control exists, the ordered pair $(\mathbf{v}, T)$ for which $\mathbf{x}_0 = \mathbf{M}(\mathbf{v}, T)$ will yield that optimal control via Equation 4.10. It is well known that the time-optimal control exists for completely controllable linear plants. It remains, then, to discover if **M** is one-to-one and onto. In order to do this, a set D is defined that has the property such that if $(\mathbf{v}, T) \in D$, then the control $u_{(0,T]} = (u(\tau)$: $u(\tau) = -\text{sgn}(\langle \mathbf{v}, \mathbf{Q}e^{\mathbf{A}\tau}\mathbf{b}\rangle)$, and $0 < \tau \leq T)$ is an optimal control for problem Z_1 and for some state $\mathbf{x} \in C/S$. One must then determine if and when $D = \partial S \times R_1^+$, etc.

Definition 4.2 (D)

Let $D \subseteq \partial S \times R_1^+$ be the set of ordered pairs in $\partial S \times R_1^+$ such that if $(\mathbf{v}, T) \in D$, then $\mathbf{v} \in \partial S$ is the terminal state at time $t = T$ of a time-optimal trajectory for some state $\mathbf{x}$ in C/S.

Definition 4.2 is of little use as it stands in writing a program to solve Problem Z_1. What is really needed is a quick way of finding out whether a given element $(\mathbf{v}, T) \in \partial S \times R_1^+$ is in D. Or better still, one might ask the question: When is it true that $D = \partial S \times R_1^+$? An example will best illustrate these points.

Example 4.1 Let the plant be the double integrator $\ddot{x}(t) = u(t)$, and let $x_1(t) = x(t)$ and $x_2(t) = \dot{x}(t)$. Then, we have

$$\begin{bmatrix} \dot{x}_1(t) \\ \dot{x}_2(t) \end{bmatrix} = \begin{bmatrix} 0 & 1 \\ 0 & 0 \end{bmatrix} \begin{bmatrix} x_1(t) \\ x_2(t) \end{bmatrix} + \begin{bmatrix} 0 \\ 1 \end{bmatrix} u(t). \tag{4.12}$$

Finally, let **Q** be the 2×2 identity matrix, and let the radius of the target be $r > 1$. In solving this problem, one finds that there are sets of points on the target on which no optimal trajectory terminates (see Figure 4.1). More formally, let $X(\mathbf{v}, T)$ denote the set

$$x(\mathbf{v}, T) = (\mathbf{x}: \mathbf{x} = \mathbf{M}(\mathbf{v}, \delta), \quad 0 < \delta \leq T). \tag{4.13}$$

There are three cases.

K1. $X(\mathbf{v}, T)$ is an optimal trajectory for all $T \in R_1^+$.

K2. $X(\mathbf{v}, T)$ is not an optimal trajectory.

K3. $X(\mathbf{v}, T)$ is an optimal trajectory if $0 < T \leq \rho(\mathbf{v}), \rho(\mathbf{v}) > 0$.

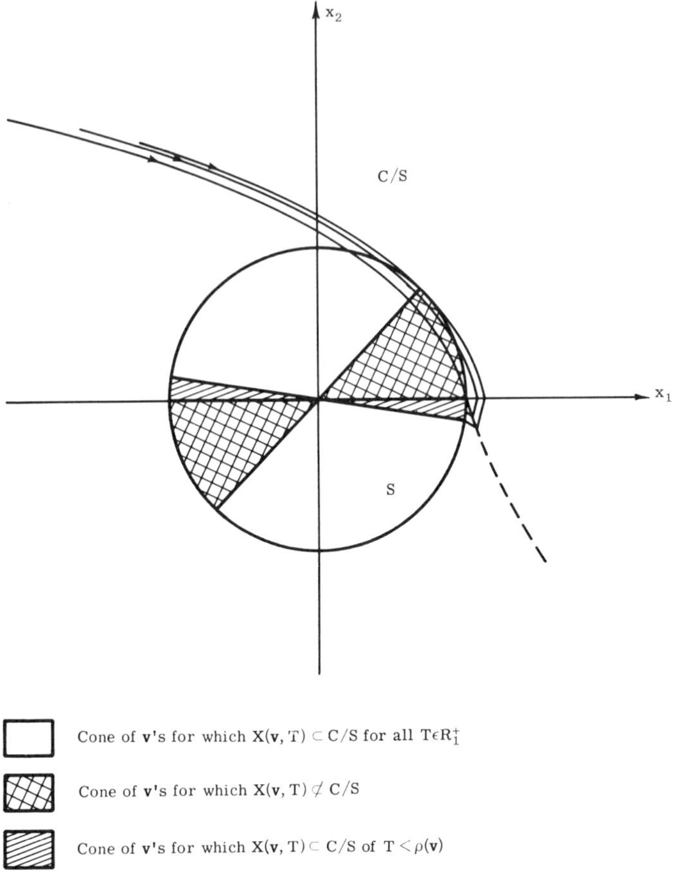

Figure 4. 1 Time-optimal trajectories to S for $r > 1$ and $\ddot{x} = u$.

It is easy to see that if $D \neq \partial S \times R_1^+$, the problem of testing only elements in D will be prohibitive. Consequently, it is important to know when $D = \partial S \times R_1^+$.

Theorem 4. 1

If all of the eigenvalues of **A** have negative real parts and if the positive-definite symmetric matrix **Q** defining the target set S is such that $\frac{1}{2}\langle \mathbf{x}, \mathbf{Qx} \rangle$ is a Liapunov function for the system

$$\dot{x}(t) = \mathbf{Ax}(t), \tag{4.14}$$

then $D = \partial S \times R_1^+$.

which contradicts Equations 4.16, 4.20, and 4.21, unless $\mathbf{v}_1 = \mathbf{v}_2$ and $T_1 = T_2$.

Now, the function $\mathbf{M}(\mathbf{v}, T)$ was defined in such a way that it automatically satisfied all of the conditions of the Minimum Principle (Theorem 1.2), except NC1 and NC9. These conditions are time-invariant and, as such, can be tested at $t = T$. Using NC1, NC9, and Equations 4.3, 4.5, 4.7, and 4.8, we find that $P_0 \geq 0$ if and only if

$$- \langle \mathbf{v}, \mathbf{QAv} \rangle + |\langle \mathbf{v}, \mathbf{Qb} \rangle| \geq 0, \tag{4.23}$$

which must hold if Equation 4.15 holds. It remains to be shown that $X(\mathbf{v}, T) \subset C/S$ for all $(\mathbf{v}, T) \in \partial S \times R_1^+$. For a given T, let $\gamma(T) \subset R_n/S = \{\mathbf{x}; \mathbf{x} = \mathbf{M}(\mathbf{v}, T), \mathbf{v} \in \partial S\}$, then the proof is completed if it can be shown that $\gamma(T) \subset R_n S$ for all $T \in R_1^+$. But the left-hand side of Equation 4.23 is precisely $(d/d\lambda) \langle x(\tau), Qx(\tau) \rangle |_{\tau=0}$, where τ is reverse time and, since it is positive (Equation 4.15) for all $\mathbf{v} \in \partial S$, $\gamma(0+) \subset R_n/S$ and, by induction on T, $\gamma(T) \subset R_n/S$ for $T \in R_1^+$. Q.E.D.

The fact that one must choose $\mathbf{Q}$ such that $\langle \mathbf{x}, \mathbf{Qx} \rangle$ is a Lyapunov function is of little consequence from the engineering point of view, since it is always possible to choose a small enough r so that the target approximates the origin for the purpose at hand. Also for many systems there will be considerable freedom in the choice of $\mathbf{Q}$. It can be shown that choosing a matrix $\mathbf{Q}$ is equivalent to choosing a particular state-space representation in which the target S is a hypersphere rather than a hyperellipse.

It can be easily seen that the conditions of Theorem 4.1 are only sufficient. For example it can be shown that $D = \partial S \times R_1^+$ in Example 4.1 if $r \leq 1$, even though $p_0 = 0$, when $v_2 = 0$. Also, if

$$\mathbf{A} = \begin{bmatrix} 0 & \omega \\ -\omega & 0 \end{bmatrix} \quad \text{and} \quad \mathbf{b} = \begin{bmatrix} 0 \\ 1 \end{bmatrix}$$

(a harmonic oscillator), then $D = \partial S \times R_1^+$ for all $r > 0$ (see Figure 30 [38]). The difficulty in finding the complete conditions under which $D = \partial S \times R_1^+$ is that Equation 4.16 will not hold if an eigenvalue of $\mathbf{A}$ has a zero or a positive real part. It is conjectured here that $D = \partial S \times R_1^+$, provided that r is small enough and the eigenvalues with zero real part are of low enough multiplicity.

Theorem 4.2

The function $\mathbf{M}$ (Definition 3.1) is continuous in $\mathbf{v}$ and T.

Proof:

For any given $\mathbf{v} \neq \mathbf{0}$, $\mathbf{M}$, considered as a function of T, is the solution of a linear-constant ordinary vector differential equation (Equation 4.9) for a piecewise constant input $u(\tau)$. Then, $\mathbf{M}$ is continuous in T.

Proof:

Since $\langle \mathbf{x}, \mathbf{Qx} \rangle$ is a Liapunov function for Equation 4.14, we s

$$\frac{1}{2} \frac{d}{dt} \langle \mathbf{x}(t), \mathbf{Qx}(t) \rangle = \langle \mathbf{x}(t), \mathbf{QAx}(t) \rangle < 0, \qquad \text{for all } \mathbf{x}(t) \in$$

Further, if $T > 0$ and $\langle \mathbf{v}_i, \mathbf{Qv}_i \rangle = r^2, i = 1, 2$, then we see tha

$$\langle \mathbf{v}_1, \mathbf{Qv}_2 \rangle \leqslant \langle \mathbf{v}_1, \mathbf{Qv}_1 \rangle,$$

where the equality holds only if $\mathbf{v}_1 = \mathbf{v}_2$ and $T = 0$.

The proof of Theorem 4.1 now takes the following form. It i shown, using Equation 4.16, that the extremal controls are u This fact, then, implies that the conditions of the Minimum F are sufficient as well as necessary. The last step is to show Equation 4.15 ensures that the conditions are met and that a sets $X(\mathbf{v}, T)$ are in C/S.

Suppose that there are two extremal controls which steer th from $\mathbf{x}_0$ to $\mathbf{v}_1 \in \partial S$ at T_1 and $\mathbf{v}_2$ at T_2, respectively, with T_1 Then, we have

$$\mathbf{v}_i = e^{\mathbf{A}T_i}\left(\mathbf{x}_0 + \int_0^T e^{-\mathbf{A}s} \mathbf{b} u_i(s) ds \right), \qquad i = 1, 2.$$

(Note that $t_0 = 0$, since the plant is assumed time-invariant.) we have

$$e^{-\mathbf{A}T_1} \mathbf{v}_1 - \int_0^{T_1} e^{-\mathbf{A}s} \mathbf{b} u_1(s)\, ds = e^{-\mathbf{A}T_2} \mathbf{v}_2 - \int_0^{T_2} e^{-\mathbf{A}s} \mathbf{b} u_2($$

But

$$u_1(s) = -\operatorname{sgn}(\langle \mathbf{v}_1, \mathbf{Q}e^{\mathbf{A}(T_1 - s)} \mathbf{b} \rangle),$$

and hence

$$-\int_0^{T_1} \langle \mathbf{v}_1, \mathbf{Q}e^{\mathbf{A}(T_1 - s)} \mathbf{b} \rangle u_1(s)\, ds \geqslant -\int_0^{T_2} \langle \mathbf{v}_1, \mathbf{Q}e^{\mathbf{A}(T_1 - s)} \mathbf{b} \rangle$$

and

$$-\int_{T_2}^{T_1} \langle \mathbf{v}_1, \mathbf{Q}e^{\mathbf{A}(T_1 - s)} \mathbf{b} \rangle u_1(s)\, ds = \int_{T_2}^{T_1} |\langle \mathbf{v}_1, \mathbf{Q}e^{\mathbf{A}(T_1 - s)} \mathbf{b} \rangle|$$

On the other hand, Equation 4.18 implies that

$$\langle \mathbf{v}_1, \mathbf{Qv}_1 \rangle - \int_0^T \langle \mathbf{v}_1, \mathbf{Q}e^{\mathbf{A}(T_1 - s)} \mathbf{b} \rangle u_1(s)\, ds$$

$$- \int_0^{T_2} \langle \mathbf{v}_1, \mathbf{Q}e^{\mathbf{A}(T_1 - S)} \mathbf{b} \rangle\, u_1(s)\, ds$$

$$= \langle \mathbf{v}_1 \mathbf{Q}e^{\mathbf{A}(T_1 - T_2)} \mathbf{v}_2 \rangle - \int_0^{T_2} \langle \mathbf{v}_1 \mathbf{Q}e^{\mathbf{A}(T_1 - S)} \mathbf{b} \rangle\, u_2(s)\, ds,$$

Also $\mathbf{M}$ can be written as the sum of two functions $\mathbf{M}_1$ and $\mathbf{M}_2$, where

$$\mathbf{M}_1(\mathbf{v}, T) = e^{-\mathbf{A}T}\mathbf{v} \tag{4.24}$$

is obviously continuous in $\mathbf{v}$, and

$$\mathbf{M}_2(\mathbf{v}, T) = e^{-AT} \int_0^T e^{\mathbf{A}s} b \operatorname{sgn}\{\langle \mathbf{v}, \mathbf{Q}e^{\mathbf{A}s}\mathbf{b}\rangle\} ds. \tag{4.25}$$

It has been shown ([34] Section V) that $\mathbf{M}_2$ is continuous in $\mathbf{v}$. Then, $\mathbf{M}$ is the sum of two continuous functions.

Q.E.D.

Theorem 4.3

If Z_1 is normal and both $\mathbf{A}$ and $\mathbf{Q}$ satisfy the conditions of Theorem 4.1, then $\mathbf{M}$ is one-to-one on D and $\mathbf{M}[D]$ is onto C/S.

Proof:

Since $\mathbf{x}_0 \in C/S$, there exists a control sequence which steers the state from $\mathbf{x}_0$ to S in finite time. It follows ([38]) Theorem 1.3) that there also exists a time-optimal control which steers the state from $\mathbf{x}_0$ to S. Hence, $\mathbf{M}[D] = C/S$. The fact that $\mathbf{M}$ is one-to-one on D can be shown in several ways. It follows almost immediately from the fact that $u^*_{(0,T^*]}$ is unique ([17] Theorem 1) for Problem Z_1. It can also be shown by assuming the contrary and demonstrating a contradiction in exactly the same manner as was used to show uniqueness of the time-optimal control for the unmodified time-optimal regulator problem in Theorem 6.7 of [3]. A geometric proof based on some results of [30] and [17] is given here because it shows why $\mathbf{M}$ is not one-to-one when the origin is the target and it also shows that $\mathbf{M}$ can be one-to-one in Problem Z_1 if the plant is nonlinear. Let $\psi(T, \mathbf{x}_0) \subset R_n$ be the set

$$\psi(T, \mathbf{x}_0) = \{\mathbf{x} : \mathbf{x} = e^{\mathbf{A}T} (\mathbf{x}_0 + \int_0^T e^{-\mathbf{A}s} bu(s)\, ds),\ |u| \leq 1\}. \tag{4.26}$$

Properties of $\psi(T, \mathbf{x}_0)$

1. $\psi(T, \mathbf{x}_0)$ is the set of all states which are reachable from $\mathbf{x}_0$ in time T (see Figure 4.2).
2. $\psi(T, \mathbf{x}_0)$ is compact and convex [30].
3. The boundary $\partial\psi(T, \mathbf{x}_0)$ of $\psi(T, \mathbf{x}_0)$ is the set [30]

$$\partial\psi(T, \mathbf{x}_0) = \{\mathbf{x} : \mathbf{x} = \mathbf{e}^{\mathbf{A}T}\{ x_0 - \int_0^T e^{-\mathbf{A}} \mathbf{b} \operatorname{sgn}\{\langle \mathbf{p}(s), \mathbf{b}\rangle\}\ ds, \dot{p}(s)$$
$$= -\mathbf{A}'\mathbf{p}(s), p(0) \neq \mathbf{0}\}. \tag{4.27}$$

4. Let $\mathbf{p}(T) = \mathbf{Q}\mathbf{v}$, $\|\mathbf{v}\|_{\mathbf{Q}} = \mathbf{r}$, and let[2]

$$\mathbf{x}(\mathbf{v}, T) = e^{\mathbf{A}T}\mathbf{x}_0 - \int_0^T e^{\mathbf{A}(T-s)}\mathbf{b} \operatorname{sgn}\{\langle \mathbf{v}, \mathbf{Q}e^{\mathbf{A}(s-T)}\mathbf{b}\rangle\}\ \mathbf{ds}, \tag{4.28}$$

[2] $\|\mathbf{v}\|_{\mathbf{Q}} = (\langle \mathbf{v}, \mathbf{Q}\mathbf{v}\rangle)^{1/2}$.

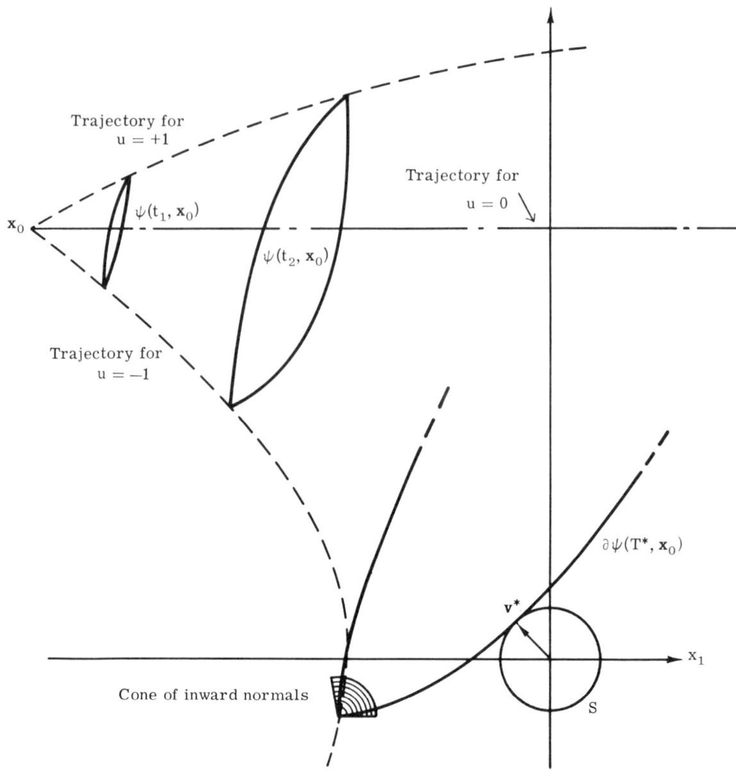

Figure 4. 2 Evolution of $\psi(t, x_0)$.

then we have

$$\langle \mathbf{v}, \mathbf{Q}(\mathbf{x}(\mathbf{v}, T) - e^{\mathbf{A}T}\mathbf{x}_0)\rangle \leqslant \langle \mathbf{v}, \mathbf{Q}(\mathbf{y} - e^{\mathbf{A}T}\mathbf{x}_0)\rangle \tag{4.29}$$

for all $\mathbf{y} \in \psi(T, \mathbf{x}_0)$, and the equality holds only if $\mathbf{y} = \mathbf{x}(\mathbf{v}, T)$ ([17] Lemma 1. Actually Equation 4.29 follows from Property 2).

5. $\psi(T, \mathbf{0})$ is continuously growing in T [25]. That is, if $T_1 > T_2$, then

$$\psi(T_1, 0) \supset \psi(T_2, 0).$$

The set $\psi(T, \mathbf{x}_0)$ can be visualized as the continuously expanding set $\psi(T, \mathbf{0})$ being translated along the undriven response from $\mathbf{x}_0$ (that is, $e^{\mathbf{A}T}\mathbf{x}_0$) of the plant. The leading edges, so to speak, contain the terminal points of the time-optimal trajectories from $\mathbf{x}_0$. Property 2 and the fact that S is also convex and compact means that the minimum time T^*, the first time that the set $\psi(T, \mathbf{x}_0) \cap S$ is not empty, is well defined. Obviously, the point of first intersection of $\psi(T, \mathbf{x}_0)$ and S occurs at some $\mathbf{v}^* \in \partial\psi(T^*, \mathbf{x}_0) \cap \partial S$. Property 4 means that the vector

$\mathbf{Qv}$ that determines the time-optimal control to $\mathbf{x}(\mathbf{v}, T)$ is an inward normal to a support hyperplane of the convex set $\psi(T, \mathbf{x}_0)$ at $\mathbf{x}(\mathbf{v}, T)$. A point $\mathbf{x}(\mathbf{v}, T)$ on $\partial\psi(T, \mathbf{x}_0)$ may have many normals (see Figure 4.1), but at the point of first contact the outward normal to S must be colinear with Qv from NC10. Since S has a unique normal everywhere, $\mathbf{Qv}^*$ is unique. Obviously, T^* is unique. Hence, for each $\mathbf{x}_0$ in C/S, there corresponds a unique ordered pair $(\mathbf{v}^*, T^*)$ in D, such that $\mathbf{x}_0 = \mathbf{M}(\mathbf{v}^*, T^*)$. Q.E.D.

Since $\mathbf{M}$ is one-to-one on D, there exists a unique inverse function $\mathbf{M}^{-1}$ from C/S onto D.

When the target is the origin, $\mathbf{M}_1(\mathbf{v}, T) = 0$, and $\mathbf{M}_2(\mathbf{v}, T)$ remains unchanged, except that the vectors $\mathbf{v}$ are chosen on the boundary of the unit hypersphere S_p in the costate space. If $\mathbf{x}_0$ is such that the first intersection of $\psi(T, \mathbf{x}_0)$ and the origin occurs at a corner of $\partial\psi(T, \mathbf{x}_0)$, then any $\mathbf{v}$ in S_p having the same direction as a vector in the cone of inward normals at the corner (see Figure 4.2) will yield $\mathbf{x}_0 = \mathbf{M}_2(\mathbf{v}, T^*)$. Hence, $\mathbf{M}_2$ cannot be one-to-one on $\partial S_p \times R_1^+$.

The convexity of $\psi(T, \mathbf{x}_0)$ is sufficient but not necessary to ensure that $\mathbf{M}$ is one-to-one. It is required only that (1) $\psi(T, \mathbf{x}_0)$ is compact, (2) the equivalent (for the nonlinear case) of Properties 3 and 4 of $\psi(T, \mathbf{x}_0)$ hold, and (3) $\psi(T, \mathbf{x}_0)$ be not too concave (that is, only one point of first contact with ∂S).

Each element $(\mathbf{v}, T)$ in D defines a piecewise constant function $u_{(0,T]}$ that alternates between the levels +1 and −1 at a finite number of switch times τ_j. Such a control is completely characterized by four parameters: (1) the initial value of the control $u(\tau = 0^+)$; (2) the number of switchings μ; (3) the set $(\tau_1, \tau_2, \ldots, \tau_\mu)$ of μ ordered switch times; and (4) the duration of the control interval T. The first three of these parameters are viewed as functions on D.

Definition 4.3 ($\eta(\mathbf{v})$, initial value of $-u(\tau)$)

Let $\eta(\mathbf{v})$ be a function from D into R_1 such that

$$\eta(\mathbf{v}) = \begin{cases} \operatorname{sgn}\{\langle \mathbf{v}, \mathbf{Qb}\rangle\}, & \langle \mathbf{v}, \mathbf{Qb}\rangle \neq 0, \\ \operatorname{sgn}\{\langle \mathbf{v}, \mathbf{Q}e^{\mathbf{A}\tau=0^+}\mathbf{b}\rangle\}, & \langle \mathbf{v}, \mathbf{Qb}\rangle = 0. \end{cases} \tag{4.30}$$

Definition 4.4 (Q_η)

Let $Q_\eta \subset D$ be a set of ordered pairs $(\mathbf{v}, T)$ such that

$$\langle \mathbf{v}, \mathbf{Qb}\rangle = 0. \tag{4.31}$$

Equation 4.31 defines a hyperplane in R_n which intersects ∂S along a great circle. It follows that Q_η is a hypersurface in D of dimension $n - 1$.

Definition 4.5 ($\mu(\mathbf{v}, T)$, number of switchings)

Let $\mu(\mathbf{v}, T)$ be an integer-valued function on D, such that the control $u_{(0,T]}$ corresponding to the ordered pair $(\mathbf{v}, T)$ makes exactly $\mu(\mathbf{v}, T)$ switchings in the open interval $(0, T)$.

Definition 4.6 ($\tau_j(\mathbf{v}), j = 1, 2, \ldots, \mu(\mathbf{v}, T)$)

The switch times $\tau_j(\mathbf{v}), j = 1, 2, \ldots, \mu(\mathbf{v}, T)$, are a set of ordered real numbers, such that

1. $\tau_j(\mathbf{v}) \in (0, T),$ (4.32)

2. $\langle \mathbf{v}, \mathbf{Q}e^{\mathbf{A}\tau_j(\mathbf{v})}\mathbf{b}\rangle = 0,$ (4.33)

3. $\text{sgn}\{\mathbf{v}, \mathbf{Q}e^{\mathbf{A}\tau_j^+(\mathbf{v})}\mathbf{b}\rangle\} = - \text{sgn}\{\langle \mathbf{v}, \mathbf{Q}e^{\mathbf{A}\tau_j^-(\mathbf{v})}\mathbf{b}\rangle\}$ (4.34)

4. $\tau_j(\mathbf{v}) < \tau_{j+1}(\mathbf{v})$ for all j. (4.35)

Not all of the zeros in $(0, T)$ of $\langle \mathbf{v}, \mathbf{Q}e^{A\tau}\mathbf{b}\rangle$ are necessarily switch times. For example, $\langle \mathbf{v}, \mathbf{Q}e^{A\tau}\mathbf{b}\rangle$ may be tangent to the time axis at a zero. Hence, the reason for Equation 4.34 is clear. Suppose one is using an analogue computer and a relay to find $\text{sgn}\{\langle \mathbf{v}, \mathbf{Q}e^{A\tau}\mathbf{b}\rangle\}$. The relay will either interpret a tangent zero as two switchings very close together or as no switchings at all. A digital computer programmed with Equations 4.33 and 4.34 in mind will treat a tangent zero as if it were not a switching.

Definition 4.7 (Q_T)

Let $Q_T \subset D$ be a set of ordered pairs $(\mathbf{v}, T)$ such that $(\mathbf{v}, T)$ satisfies Equations 4.33 and 4.34.

If $(\mathbf{v}, T) \in Q_T$, then T coincides with a switch time of $u(\tau)$ at T. Let $\dot{\mathbf{x}}(\mathbf{v}, T)$ denote the reverse-time state velocity $\dot{\mathbf{x}}$ for the trajectory $X(\mathbf{v}, T)$ evaluated at $\tau = T$. Then, $\dot{\mathbf{x}}(\mathbf{v}, T)$ is not defined if $(\mathbf{v}, T) \in Q_T$, since $X(\mathbf{v}, T^*)$ has a corner at $\tau = T$. The set $Q_T \subset D$ has dimension $n - 1$.

Definition 4.8 (Q_μ)

Let $Q_\mu \subset D$ be a set of ordered pairs $(\mathbf{v}, T)$, such that for one or more times τ in $(0,T]$, the equations

$$\langle \mathbf{v}, \mathbf{Q}e^{A\tau}\mathbf{b}\rangle = 0, \tag{4.36}$$

$$\langle \mathbf{v}, \mathbf{Q}\mathbf{A}e^{\mathbf{A}\tau}\mathbf{b}\rangle = 0 \tag{4.37}$$

are satisfied.

Since the problem is normal, the vectors **b**, **Ab** are linearly independent (see Appendix C). Then, for any T_q in R_1^+ and $n = 3$, there exists exactly one vector $\mathbf{v}_q$ such that $\pm\mathbf{v}_q \in \partial S$ and $(\mathbf{v}_q, T_q)$ satisfy Equations 4.36 and 4.37. As T_q ranges over R_1^+, the $\pm\mathbf{v}_q$ vector will trace out a set in ∂S of dimension one. It follows from Definition 4.8 that Q_μ has dimention 2 (that is, if $n = 3$). If $n > 3$, one can similarly show that there is a set $Q_\mu \subset D$ of dimension $n - 1$ on which Equations 4.36 and 4.37 are satisfied. Further, the set Q_μ is comprised of subsets $Q_{\mu i}$, $i = 1, 2, \ldots, n-2$, such that if $(\mathbf{v}, T) \subset Q_{\mu i}$ then

$$\langle \mathbf{v}, \mathbf{Q}\mathbf{A}^l e^{\mathbf{A}\tau}\mathbf{b}\rangle = 0, \qquad l = 0, 1, \ldots, i, \tag{4.38}$$

and

$$\langle \mathbf{v}, \mathbf{Q}\mathbf{A}^{i+1} e^{\mathbf{A}\tau}\mathbf{b}\rangle \neq 0 \tag{4.39}$$

for some time τ in $(0, T)$. Obviously, if i in Equations 4.38 and 4.39 is odd, the time τ is a tangent zero, and, if i is even, the time τ is a switch time. It also follows that the set $Q_{\mu i} \subset D$ has dimension $n - i$.

Example 4.2 Consider the plant

$$\begin{bmatrix} \dot{x}_1 \\ \dot{x}_2 \end{bmatrix} = \begin{bmatrix} -\alpha & \beta \\ -\beta & -\alpha \end{bmatrix} \begin{bmatrix} x_1 \\ x_2 \end{bmatrix} + \begin{bmatrix} 0 \\ 1 \end{bmatrix} u, \qquad |u| \leqslant 1,$$

and let

$$x_1 = r\cos\theta, \quad r > 0, \qquad x_2 = r\sin\theta, \qquad 0 \leqslant \theta < 2\pi.$$

(That is, $\mathbf{Q} = \mathbf{I}$.)

The set $D = \partial S \times R_1^+$ is illustrated in the $\theta \sim T$ plane in Figure 4.3, which shows the sets Q_η and the set Q_T for this plant. The set Q_μ is empty, since $n < 3$.

Example 4.3. Consider the plant

$$\begin{bmatrix} \dot{x}_1 \\ \dot{x}_2 \end{bmatrix} = \begin{bmatrix} 0 & 1 \\ 0 & 0 \end{bmatrix} \begin{bmatrix} x_1 \\ x_2 \end{bmatrix} + \begin{bmatrix} 0 \\ 1 \end{bmatrix} u, \qquad |u| \leqslant 1,$$

and let $r \leqslant 1$. The set D for this plant is illustrated in the $\theta \sim T$ plane in Figure 4.4, which shows the set Q_η and Q_T for this plant.

Definition 4.9 (Q)

Let Q denote the union

$$Q = Q_\mu \cup Q_T \cup Q_\eta. \tag{4.40}$$

That is, Q is a collection of $n-1$ dimensional hypersurfaces in D.

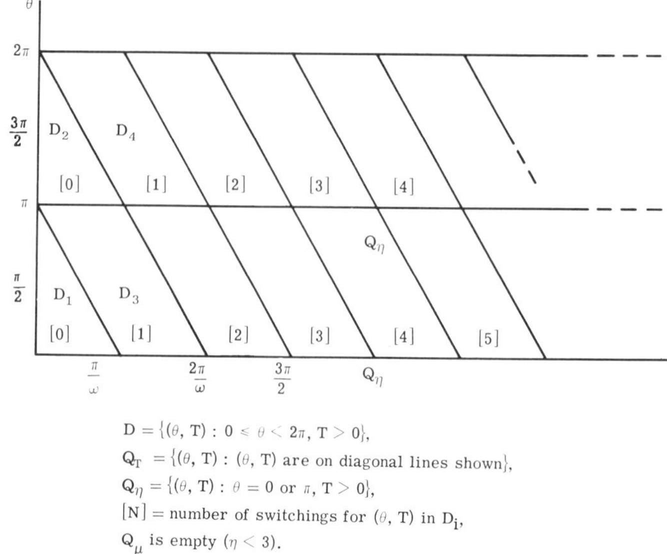

Figure 4. 3 The set D for an oscillator with or without damping.

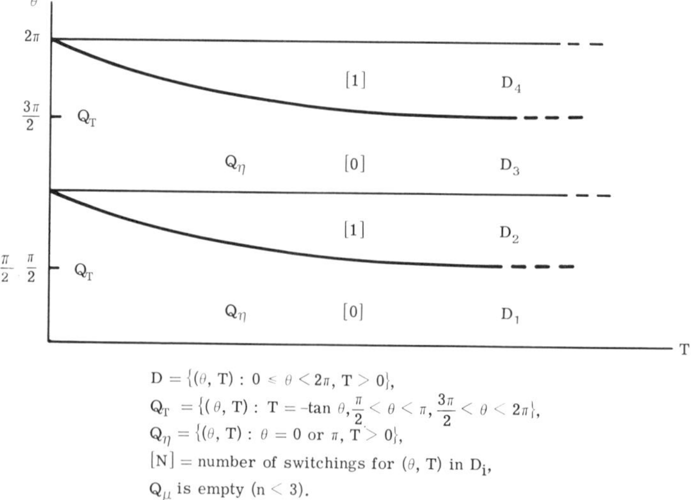

Figure 4. 4 The set D for the double integrator plant when r $\leq$ 1.

Definition 4.10 (D_i)

Let D_i, $i = 1, 2, \ldots,$ denote the ordering of the disjoint open subsets comprising D/Q. That is,

$$D/Q = \bigcup_{i=1,2,\ldots,} D_i$$

The fact that Q is composed of a network of (n − 1)-dimensional hypersurfaces that divide D into a collection of open subsets D_i means that given a point $(\mathbf{v}_q, T_q)$ and an $\epsilon > 0$, there exists a neighborhood in D of the point $(\mathbf{v}_q, T_q)$ of radius ϵ containing infinitely many points in two or more of the subsets D_i as well as infinitely many points in Q.

The principal reason for classifying these sets of points making up Q was to prepare the way for obtaining the derivatives of **M** in D. These derivatives are obtained by introducing a hyperspherical coordinate transformation.

Definition 4.11 (Hyperspherical coordinates)

Let Θ be a subset of R_{n-1} of $n - 1$ vectors θ whose components θ_i, $i = 1, 2, \ldots, n - 1$, satisfy the relations

$$0 \leqslant \theta_i \leqslant \pi, \qquad i = 1, 2, \ldots, n - 2, \tag{4.41}$$

$$0 \leqslant \theta_{n-1} < 2\pi, \tag{4.42}$$

and let $\mathbf{h}(\theta)$ be a function from Θ into R_n, where

$$h_1(\theta) = \cos \theta_1, \tag{4.43}$$

$$h_k(\theta) = \cos \theta_k \prod_{j=1}^{k-1} \sin \theta_j \qquad (k = 2, 3, \ldots, n - 1), \tag{4.44}$$

$$h_n(\theta) = \prod_{j=1}^{n-1} \sin \theta_j. \tag{4.45}$$

Properties of h(θ)

1. $\|\mathbf{h}(\theta)\| = 1$.
2. $\mathbf{h}[\Theta] = \partial S_1$, where S_1 is the unit hypersphere.
3. $\mathbf{h}(\theta)$ is one-to-one on Θ except if $\theta_i = 0$ or π, $i = 1, 2, \ldots, n - 2$. The corresponding $2(n - 2)$ points on ∂S_1 are called the poles of the hypersphere.
4. $\mathbf{h}(\theta)$ has continuous partial derivatives of all order everywhere. Properties 1 through 4 follow immediately from Equations 4.41 through 4.45.

5. If $\mathbf{v} \in \partial S$ (that is, $\langle \mathbf{v}, \mathbf{Qv} \rangle = r^2$), then $\mathbf{v} = r\mathbf{L}^{-1}\mathbf{h}(\theta)$, where $\mathbf{L}$ is the unique triangular matrix such that $\mathbf{Q} = \mathbf{L}'\mathbf{L}$.

Property 5 follows, since every nonsingular symmetric matrix can be written as a product $\mathbf{Q} = \mathbf{L}'\,\mathbf{L}$, where $\mathbf{L}$ is triangular and nonsingular. ([16] p. 37, Vol. I). Then, $\langle \mathbf{v}, \mathbf{Qv} \rangle = \langle \mathbf{Lv}, \mathbf{Lv} \rangle = r^2$, and, since 1 holds, $\mathbf{Lv} = r\mathbf{h}(\theta)$ or $\mathbf{v} = r\mathbf{L}^{-1}\mathbf{h}(\theta)$.

Let $\mathbf{h}_\theta(\theta)$ denote the first derivative of $\mathbf{h}(\theta)$ with respect to $\boldsymbol{\theta}$. That is,

$$h_\theta(\theta) = \begin{bmatrix} \dfrac{\partial h_1(\theta)}{\partial \theta_1} & \dfrac{\partial h_1(\theta)}{\partial \theta_2} & \cdots & \dfrac{\partial h_1(\theta)}{\partial \theta_{n-1}} \\ \vdots & & & \vdots \\ \dfrac{\partial h_n(\theta)}{\partial \theta_1} & \dfrac{\partial h_n(\theta)}{\partial \theta_2} & & \dfrac{\partial h_n(\theta)}{\partial \theta_{n-1}} \end{bmatrix} \tag{4.46}$$

is an $n \times (n-1)$ matrix.

Properties of $\mathbf{h}_\theta(\theta)$ (see Appendix E)

1. The $n \times (n-1)$ matrix $\mathbf{h}_\theta(\theta)$ exists everywhere in Θ and has mutually orthogonal nonzero column vectors everywhere on Θ, except at a pole of the hypersphere (that is, $\theta_i = 0$ or π, $i = 1, 2, \ldots, n-2$). At a pole, one or more of the column vectors of $\mathbf{h}_\theta(\theta)$ is the zero vector.
2. The n-vector $\mathbf{h}'_\theta(\theta)\mathbf{h}(\theta) = 0$ for all θ in Θ.

Property 2 of $\mathbf{h}(\theta)$ means that given a $\mathbf{v} \in \partial S$, there exists a unique $\theta = (1/r)\mathbf{h}^{-1}(\mathbf{Lv})$ unless $\mathbf{Lv}$ is a pole of S. However, the poles are arbitrary in that one can redefine the relationship between the v_i and the $h_i(\theta)$ and place the poles anywhere relative to the Cartesian coordinates of S. Since $\mathbf{h}(\theta)$ is used merely to find the derivative of $\mathbf{M}$ when restricted to D, one can always redefine the $h_i(\theta)$ if the current guess is too close to a pole. It is henceforth assumed that one can always write $\theta = (1/r)\mathbf{h}^{-1}(\mathbf{Lv})$.

Theorem 4.4

The function $\mathbf{M}(\mathrm{v}, \mathrm{T})$ has continuous partial derivatives of all order on every open set D_i.

Proof:

Let $(\mathbf{v}, T) \in D_i$. Then, on D_i, $\mathbf{M}$ has a representation

$$\mathbf{M}(\mathrm{v}, \mathrm{T}) = e^{-\mathbf{A}T}\left(\mathrm{v} + \sum_{i=1}^{\mu} (-1)^j \int_{\tau_j(\mathbf{v})}^{\tau_{j+1}(\mathbf{v})} e^{\mathbf{A}s} b \, ds\right), \tag{4.47}$$

where $\tau_0 = 0$ and $\tau_{\mu+1}(\mathbf{v}) = T$. The derivatives of **M** with respect to **v** must be restricted to the set D. This is achieved by taking the derivative of **M** with respect to θ, where

$$\mathbf{v} = r\mathbf{L}^{-1}\mathbf{h}(\theta). \tag{4.48}$$

Here $\mathbf{M}_1(v, T)$ (see Equation 4.24) clearly has continuous partial derivatives of all orders with respect to the components θ_i, $i = 1, 2, \ldots, n-1$ and T, in view of Equation 4.48 and Property 4 of $\mathbf{h}(\theta)$. The function $\mathbf{M}_2(\mathbf{v}, T)$ (see Equation 4.25) is a function of θ indirectly through $\tau_j(\mathbf{v})$ and Equation 4.48. Hence, using the chain rule for differentiation, we obtain

$$\frac{\partial \mathbf{M}_2}{\partial \theta_i} = \sum_{j=1}^{\mu} \sum_{k=1}^{n} \frac{\partial \mathbf{M}}{\partial \tau_j} \frac{\partial \tau_j}{\partial v_k} \frac{\partial v_k}{\partial \theta_i} \tag{4.49}$$

and

$$\frac{\partial \mathbf{M}_2}{\partial \tau_j} = 2\eta(-1)^{j+1} e^{\mathbf{A}(\tau_j(\mathbf{v}) - T)}\mathbf{b}, \tag{4.50}$$

which is continuous and has continuous partial derivatives of all order with respect to $\tau_j(\mathbf{v})$. Since $\mu(\mathbf{v}, T)$ and $\eta(\mathbf{v})$ are constant on D_i, the implicit function

$$\langle \mathbf{v}, \mathbf{Q}e^{\mathbf{A}\tau_j(\mathbf{v})}\mathbf{b}\rangle = 0 \tag{4.51}$$

has μ solutions everywhere on D_i. By taking the total derivative of Equation 4.51, we obtain

$$\frac{\partial \tau_j}{\partial v_i} = -\frac{\langle \mathbf{i}, \mathbf{Q}\, e^{\mathbf{A}\tau_j(\mathbf{v})}\mathbf{b}\rangle}{\langle \mathbf{v}, \mathbf{Q}, \mathbf{A}e^{\mathbf{A}\tau_j(\mathbf{v})}\mathbf{b}\rangle}, \qquad i = 1, 2, \ldots, n, \qquad j = 1, 2 \ldots, \mu, \tag{4.52}$$

where i is an n-vector whose ith element is 1 and whose other elements are all zero. It follows from the definition of D_i that $\partial \tau_j / \partial v_i$ exists. Furthermore, $\partial \tau_j / \partial v_i$ is continuous on D_i, and $\partial \tau_j / \partial v_i$ has continuous partial derivatives of all orders. The higher-order partial derivatives may be similarly obtained. Now, we have

$$\begin{aligned} \frac{\partial \mathbf{M}_2}{\partial T} &= -\mathbf{A}\,\mathbf{M}_2(\mathbf{v}, T) + e^{-\mathbf{A}T}\,\eta(-1)^{\mu+1} e^{\mathbf{A}T}\mathbf{b} \\ &= -\mathbf{A}\,\mathbf{M}_2(\mathbf{v}, T) + \eta(-1)^{\mu+1}\mathbf{b}, \end{aligned} \tag{4.53}$$

and

$$\frac{\partial^2 \mathbf{M}_2}{\partial T \partial \theta_i} = \mathbf{0}. \tag{4.54}$$

Hence, the theorem is proved.

Definition 4.12 (G)

Let $(\mathbf{v}, T) \in D_i$ (for some i) and let $\boldsymbol{\theta} = (1/r)\mathbf{h}^{-1}(\mathbf{Lv})$. Then, the first derivative of **M**, in the sense of Frechet (see Appendix B and [15] p. 190) restricted to D and with respect to all arguments, is an n × n matrix denoted by $\mathbf{G}(\mathbf{v}, T)$, or, simply, by **G**. Thus, we have

$$G(\mathbf{v}, T) = [r\mathbf{M}_{\mathbf{v}}(\mathbf{v}, T)\, \mathbf{L}^{-1}\mathbf{h}_{\theta}(\theta) \mid \mathbf{M}_T(\mathbf{v}, T)], \tag{4.55}$$

$$\theta = (1/r)\mathbf{h}^{-1}(\mathbf{Lv}), \tag{4.56}$$

where $\mathbf{M}_T(\mathbf{v}, \mathbf{T})$ is the partial derivative of **M** with respect to T at $(\mathbf{v}, T)$, and $M_{\mathbf{v}}(\mathbf{v}, T)$ is the n × n matrix

$$\mathbf{M}_{\mathbf{v}}(\mathbf{v}, T) = \left[\frac{\partial \mathbf{M}(\mathbf{v}, T)}{\partial v_1} \; \frac{\partial M(\mathbf{v}, T)}{\partial v_2} \cdots \frac{\partial \mathbf{M}(\mathbf{v}, T)}{\partial v_n} \right]. \tag{4.57}$$

Theorem 4.5

Let $(\mathbf{v}, T) \in D_i$ (for some i). Then, the n × n matrix $\mathbf{G}(\mathbf{v}, T)$ has rank n if and only if

$$-\langle \mathbf{v}, \mathbf{QAv}\rangle + |\langle \mathbf{b}, \mathbf{Qv}\rangle| > 0. \tag{4.58}$$

Proof:

From Equation 4.47, we have

$$\mathbf{M}_{\mathbf{v}}(\mathbf{v}, T) = e^{-\mathbf{A}T} \left\{ \mathbf{I} + \sum_{j=0}^{\mu} \eta(-1)^j \left\{ e^{\mathbf{A}\tau_{j+1}(\mathbf{v})}\mathbf{b} \rangle \langle \frac{\delta \tau_{j+1}(\mathbf{v})}{\partial \mathbf{v}} - e^{\mathbf{A}\tau_j(\mathbf{v})}\mathbf{b} \rangle \langle \frac{\delta \tau_j(\mathbf{v})}{\partial \mathbf{v}} \right\} \right\}. \tag{4.59}$$

From Equation 4.51, and from the fact that $\tau_0 = 0$ and $\tau_{\mu+1} = T$, we see that

$$\frac{\partial \tau_j}{\partial \mathbf{v}} = \frac{-1}{\langle \mathbf{v}, \mathbf{QA}\, e^{\mathbf{A}\tau_j}\mathbf{b}\rangle} \mathbf{Q}\, e^{\mathbf{A}\tau_j}\mathbf{b}, \qquad j = 1, 2, \ldots, \mu, \tag{4.60}$$

and

$$\frac{\partial \tau_0}{\partial \mathbf{v}} = \frac{\partial \tau_{\mu+1}}{\partial \mathbf{v}} = 0. \tag{4.61}$$

Collecting terms in Equation 4.59 and using Equations 4.60 and 4.61, we have

$$\mathbf{M_v}(\mathbf{v}, T) = e^{-\mathbf{A}T} \{\mathbf{I} + \mathbf{E}\}, \tag{4.62}$$

where $\mathbf{I}$ is the $n \times n$ identity matrix, and

$$\mathbf{E} = \sum_{j=1}^{\mu} \frac{2\eta(-1)^{j+1}}{\langle \mathbf{v}, \mathbf{Q}\mathbf{A}e^{\mathbf{A}\tau_j}\mathbf{b}\rangle} e^{\mathbf{A}\tau_j}\mathbf{b}\rangle\langle \mathbf{Q}e^{\mathbf{A}\tau_j}\mathbf{b}. \tag{4.63}$$

Since the controls alternate in sign, and since $(\mathbf{v}, T) \notin \mathbf{Q}$, we have

$$\operatorname{sgn} \{\langle \mathbf{v}, \mathbf{Q}\mathbf{A}e^{\mathbf{A}\tau_j}\mathbf{b}\rangle\} = \eta(-1)^{j+1}. \tag{4.64}$$

Hence, we have

$$\mathbf{E} = \sum_{j=1}^{\mu} \frac{2}{|\langle \mathbf{v}, \mathbf{Q}\mathbf{A}e^{\mathbf{A}\tau_j}\mathbf{b}\rangle|} e^{\mathbf{A}\tau_j}\mathbf{b}\rangle\langle \mathbf{Q}e^{\mathbf{A}\tau_j}\mathbf{b}. \tag{4.65}$$

Obviously, since $\mathbf{Q} = \mathbf{L}'\mathbf{L}$,

$$\langle \mathbf{x}, \mathbf{E}\mathbf{x}\rangle \geq 0, \qquad \text{for all } \mathbf{x} \in R_n. \tag{4.66}$$

And, since τ_j is a switch time,

$$\langle \mathbf{v}, \mathbf{E}\mathbf{v}\rangle = 0. \tag{4.67}$$

From Equations 4.66 and 4.67 it is concluded that $\mathbf{E}$ is positive semidefinite. Then, $\mathbf{I} + \mathbf{E}$ is positive definite and hence is nonsingular. The matrix $e^{-\mathbf{A}T}$ is nonsingular for all $\mathbf{A}$ and all finite T, since it is a fundamental matrix (see Appendix B). Then, the matrix $e^{-\mathbf{A}T}[\mathbf{I} + \mathbf{E}]$ has rank n (that is, is nonsingular). In Equation 4.55, the $n \times (n-1)$ matrix $\mathbf{h}_\theta(\theta)$, $\theta = (1/r)\mathbf{h}^{-1}(\mathbf{L}\mathbf{v})$, has $n-1$ mutually orthogonal nonzero column vectors (and hence they are linearly independent). Then, the $n \times (n-1)$ matrix $e^{-\mathbf{A}T}[\mathbf{I} + \mathbf{E}]\mathbf{L}^{-1}\mathbf{h}_\theta(\theta)$ has rank $n-1$. Now, we have

$$\mathbf{G} = e^{-\mathbf{A}T}[\mathbf{I} + \mathbf{E}]\mathbf{L}^{-1}\mathbf{h}_\theta(\theta)\mathbf{M}_T. \tag{4.68}$$

Hence, $\mathbf{G}(\mathbf{v}, T)$ will have rank n if and only if the n-vector $\mathbf{M}_T(\mathbf{v}, T)$ is linearly independent of the other $n-1$ columns of $\mathbf{G}(\mathbf{v}, T)$. From Property 2 of $\mathbf{h}_\theta(\theta)$ and Equations 4.51, 4.65, and 4.48, it follows that the vector $\mathbf{p}(T) = e^{\mathbf{A}'T}\mathbf{Q}\mathbf{v}(\mathbf{v} = r\mathbf{L}^{-1}\mathbf{h}(\theta))$ is orthogonal to each of the first $n-1$ columns of $\mathbf{G}(\mathbf{v}, T)$; that is,

$$[e^{-\mathbf{A}T}[\mathbf{I} + \mathbf{E}]\mathbf{L}^{-1}\mathbf{h}_\theta(\theta)]' e^{\mathbf{A}'T}\mathbf{Q}\mathbf{v} = \mathbf{0}. \tag{4.69}$$

Hence, $\mathbf{G}(\mathbf{v}, T)$ will have rank n if and only if $\langle e^{\mathbf{A}'T}\mathbf{Q}\mathbf{v}, \mathbf{M}_T(\mathbf{v}, T)\rangle \neq 0$. But $\mathbf{M}_T(\mathbf{v}, T) = \dot{\mathbf{x}}(\mathbf{v}, T)$, and the scalar product $\langle \mathbf{p}(\tau), \dot{\mathbf{x}}(\mathbf{v}, T)\rangle$ is constant along an optimal trajectory. From NC1, NC9, and Equation 4. 3, we have

$$\langle \mathbf{p}(T), \mathbf{M}_T(\mathbf{v}, T)\rangle = -\langle \mathbf{v}, \mathbf{QAv}\rangle + |\langle \mathbf{v}, \mathbf{Q}b\rangle| \geq 0.$$

Q.E.D.

It is henceforth assumed that $\mathbf{G}(\mathbf{v}, T)$ has rank n. (Note that Condition 4. 58 in Theorem 4. 5 is the same as Condition 4. 23 in Theorem 4. 1). The matrix $\mathbf{G}(\mathbf{v}, T)$ and its inverse, the matrix $\mathbf{G}^{-1}(\mathbf{v}, T)$, considered as functions from $R_n \times R_1^+$ into $R_n \times R_n$, are continuous on each open set $D_i \subset D$, and they are discontinuous on Q_T and Q_η, where the discontinuity is of the form of the addition and/or subtraction of one or more dyads of the form $(2/|\langle \mathbf{v}, \mathbf{QA}e^{\mathbf{A}\tau_j}b\rangle|)e^{\mathbf{A}\tau_j}b\rangle\langle \mathbf{Q}e^{\mathbf{A}\tau_j}b$ to E and/or a change in $\mathbf{M}_T(\mathbf{v}, T) = -\mathbf{AM}(\mathbf{v}, T) + b\eta(-1)^\mu$ caused by $\eta(-1)^\mu$ changing sign. The latter produces a smaller effect when $\|\mathbf{M}(\mathbf{v}, T)\|$ is large, and the former produces a smaller effect when μ is large and in stable systems when τ_j is large.

Let $(\mathbf{v}_q, T_q) \in Q_T \cup Q_\eta$ (that is, excluding Q_μ), then the

$$\lim_{\substack{\mathbf{v}\to\mathbf{v}_q \\ \mathbf{v}\to T_q}} \mathbf{G}(\mathbf{v}, T)$$

exists but depends on the set D_i through which the limit is taken. If $(\mathbf{v}_q, T_q) \in Q_\mu$, then the limit does not exist, since elements of $\mathbf{G}(\mathbf{v}, T)$ become arbitrarily large. Hence, the set Q_μ could be a really "bad set." It will be assumed for the remainder of the chapter that $\mathbf{Q} = \mathbf{I}$. In other words, it is assumed that a particular state-space representation has been chosen, such that the hypersphere is a Lyapunov function (Appendix D). This is done merely for convenience and clarity. In programming the iterative procedure presented in the next section, one has the choice of selecting a suitable state-space representation and using the hypersphere as a target, or of choosing a suitable $\mathbf{Q}$ matrix and leaving the state-space representation as it is initially.

4. 4 The Iterative Procedure

The iterative procedure is based on the fact that $\mathbf{G}(\mathbf{v}, T)$ and its inverse matrix $\mathbf{G}^{-1}(\mathbf{v}, T)$ exist almost everywhere in D. A guess $(\mathbf{v}, T)$ is made in D/Q, and it is tested for an error $\mathbf{e}(\mathbf{v}, T) = \mathbf{x}_0 - \mathbf{M}(\mathbf{v}, T)$. The function $\mathbf{M}$ and the constraint $\mathbf{v} \in \partial S$ are then linearized about the point $(\mathbf{v}, T)$, and this linearized approximation is inverted to find a better guess in the sense of reducing the norm of the error $\|\mathbf{e}(\mathbf{v}, T)\|$. The problem of making a good initial guess is another matter and is treated in the sequel (see Section 4. 6).

Suppose that a kth guess $(\mathbf{v}_k, T_k)$ has been made. The following recursion formulas describe the iterative procedure. (Note: it is assumed that $\mathbf{Q} = \mathbf{I}$.)

$$x_k = \mathbf{M}(v_k, T_k), \qquad (v_k, T_k) \in D/\mathbf{Q}, \tag{4.70}$$

and

$$e_k = x_0 - x_k, \tag{4.71}$$

where

$$x_0 = \mathbf{M}(v^*, T^*). \tag{4.72}$$

Let

$$\theta_k = \frac{1}{r} h^{-1}(v_k) \tag{4.73}$$

and

$$G_k = [r\mathbf{M}_v(v_k, T_k)h_\theta(\theta_k) \mid \mathbf{M}_T(v_k, T_k)]. \tag{4.74}$$

Let

$$\begin{bmatrix} \Delta\hat{\theta}_k \\ \hline \Delta\hat{T}_k \end{bmatrix} = G_k^{-1} e_k, \tag{4.75}$$

$$\Delta\hat{y}_k = h_\theta(\theta_k)\Delta\hat{\theta}_k, \qquad \text{an n-vector,} \tag{4.76}$$

$$v_{k+1}(\gamma_k) = \frac{r(v_k + \gamma_k \Delta\hat{\gamma}_k)}{\|v_k + \gamma_k \Delta\hat{\gamma}_k\|}, \tag{4.77}$$

and

$$T_{k+1}(\gamma_k) = T_k + \gamma_k \Delta\hat{T}_k. \tag{4.78}$$

Choose γ_k^* in $(0, 1]$ such that

$$\|e_{k+1}\| = \min_{0<\gamma_k\leq 1} \|x_0 - \mathbf{M}(v_{k+1}(\gamma_k), T_{k+1}(\gamma_k))\|, \tag{4.79}$$

and then choose

$$v_{k+1} = v_{k+1}(\gamma_k^*)$$

and

$$T_{k+1} = T_{k+1}(\gamma_k^*).$$

As previously stated, the role played by $\mathbf{h}(\theta)$ is strictly a means of finding the derivative of $\mathbf{M}$ restricted to D. The linearization is done in this manner for consistency, since $\mathbf{v}$ is constrained to lie on ∂S. The vector $\Delta\hat{\gamma}_k$ defined by Equation 4.76 is tangent to ∂S at $\mathbf{v}_k$ from Property 2 of $\mathbf{h}(\theta)$. The two vectors $\Delta\hat{\gamma}_k$ and $\mathbf{v}_k$ define a hyperplane which intersects the hypersphere along a great circle g_k. Equation 4.77 provides a means of selecting a $\mathbf{v}_{k+1}$ on g_k (thus ensuring that $\mathbf{v}_{k+1} \in \partial S$), and Equation 4.78 adjusts the time accordingly, within first-order terms.

A typical step in the iterative procedure is illustrated in Figure 4.5. The matrix $\mathbf{G}$ has a geometric interpretation in terms of support hyperplanes to the minimum isochrones and tangent vectors to the time-optimal trajectories. The first $n - 1$ columns of $\mathbf{G}$ represent the rate of change of $\mathbf{M}(\mathbf{v}, T)$ with respect to changes in $\mathbf{v}$ that are restricted to ∂S. Since T is fixed under such changes, the effect is to move $\mathbf{x} = \mathbf{M}(\mathbf{v}, T)$ along the minimum isochrone containing $\mathbf{x}$. Thus, the isochrone is represented at $\mathbf{M}(\mathbf{v}, T)$ by its support hyperplane at that point. The first $n - 1$ columns of $\mathbf{G}$ lie in this support hyperplane [since $\mathbf{p}(T)$, which is known to be normal to this hyperplane, is orthogonal to each of these $n - 1$ columns]. Also, $\mathbf{M}_T(\mathbf{v}, T) = \dot{\mathbf{x}}(T)$ is tangent to the time-optimal trajectory passing through $\mathbf{M}(\mathbf{v}, T)$ (see Figure 4.5). The quality of this representation (that is, approximating the isochrones by hyperplanes and the trajectories by straight lines) will depend on the local curvature of the minimum isochrones and the time-optimal trajectories. The columns of $\mathbf{G}$ (see Figure 4.5) can be viewed as defining a local skewed-coordinate system, centered at $\mathbf{M}(\mathbf{v}, T)$, in which the error vector $\mathbf{e}_k(\mathbf{v}, T)$ is measured in terms of the changes $\Delta\hat{\theta}_k$ and $\Delta\hat{T}_k$ by using Equation 4.75.

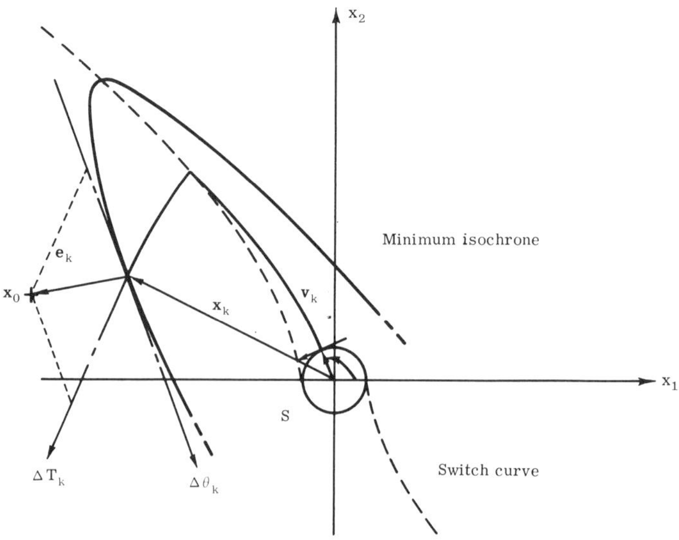

Figure 4.5 Illustration of iteration for a second-order plant.

4.5 Convergence

Three situations are considered with respect to convergence. Equations 4.77 and 4.78 describe a path $(\mathbf{v}_{k+1}(\gamma_k), T_{k+1}(\gamma_k))$ in D parameterized by γ_k along which the new error $\|e_{k+1}(\gamma_k)\|$ is evaluated. The three situations are as follows: (1) $(\mathbf{v}_k, T_k)$ and $(\mathbf{v}^*, T^*)$ are "close" together and in the same subset $D_i \subset D$, and the path $(v_{k+1}(\gamma_k), T_{k+1}(\gamma_k))$ remains in D_i; (2) the same situation as in (1), except that the points $(\mathbf{v}^*, T^*)$ and $(\mathbf{v}_k, T_k)$ are not "close" enough; and (3) the point $(\mathbf{v}^*, T^*)$ is in D_i, $(\mathbf{v}_k, T_k) \in D_j$ $(i \neq j)$, and the path encounters the set $Q_T \cup Q_\eta$ for $\gamma_k = \delta_k$.

The iterative procedure is closely related to Newton's method. In fact, if $\Delta\hat{y}_k$ and $\Delta\hat{T}_k$ are small enough, Equations 4.77 and 4.78 can be replaced by

$$\begin{bmatrix} \theta_{k+1} \\ \hline T_{k+1} \end{bmatrix} = \begin{bmatrix} \theta_k \\ \hline T_k \end{bmatrix} + \mathbf{G}_{\bar{k}}^{-1}\mathbf{e}_k , \qquad (4.80)$$

where y_k has been chosen as 1. Equation 4.80 suggests that one might consider that $\mathbf{M}$ maps $\Theta \times R_1^+$ into R_n and search in $\Theta \times R_1^+$ for the points (θ^*, T^*), such that

$$\mathbf{x}_0 = \mathbf{M}(rh\theta^*), T^*) \qquad (4.81)$$

or

$$x_0 = \tilde{M}(\theta^*, T^*). \qquad (4.82)$$

Several theorems appear in the literature concerning the convergence and the rate of convergence of Newton's method (see [31] and [39]). These can best be interpreted for Problem Z_1 in terms of $\tilde{\mathbf{M}}$. Paraphrasing a theorem from Todd [40], we let

$$\mathbf{w} = \begin{bmatrix} \theta \\ \hline T \end{bmatrix}, \qquad (4.83)$$

and

$$\tilde{D} = \Theta \times R_1^+. \qquad (4.84)$$

Theorem 4.6

Let $\tilde{\mathbf{M}}$ be a function defined on $\tilde{D} \subset \mathbf{R}_a\{\tilde{\mathbf{M}}\}$[3] in R_n and let $\mathbf{w}_0 \in \tilde{D}$ be selected, such that

---

[3] $R_a\{\mathbf{M}\}$ means the range of $\mathbf{M}$.

1. $\tilde{\mathbf{M}}$ has continuous second partial derivatives on an open neighborhood of $\mathbf{w}_0$, $N(\mathbf{w}_0, r_0) = \{\mathbf{w}: \| \mathbf{w} - \mathbf{w}_0 \| < r_0\}$, such that

$$\sum_{i=1}^{n} \sum_{j=1}^{n} \sum_{k=1}^{n} \left(\frac{\partial^2 \tilde{M}_i}{\partial w_j \partial w_k} \right)^2 \leqslant k^2 \text{ on } N(\mathbf{w}_0, r_0), \tag{4.85}$$

where $\tilde{\mathbf{M}}_i$ is the ith element of $\mathbf{M}$;

2. $|\tilde{\mathbf{M}}_w|$[4] $\neq 0$, $\| \mathbf{M}_{\bar{w}}^{-1}(\mathbf{w}_0)\mathbf{M}(\mathbf{w}_0) \| \leqslant \eta_0$, and

$$\sum_{i=1}^{n} \sum_{h=1}^{n} \left[\left(\frac{\partial w_i}{\partial \tilde{M}_k} \right) \Bigg|_{\mathbf{w}=\mathbf{w}_0} \right]^2 \leqslant \beta_0^2; \tag{4.86}$$

3. the constant $h_0 = \beta_0 \eta_0 k \leqslant 1/2$, and

$$\frac{1}{h_0}(1 - \sqrt{1 - 2h_0})\eta_0 < r_0. \tag{4.87}$$

Then, the sequence $\{\mathbf{w}_k\}$ defined by

$$\mathbf{w}_{k+1} = \mathbf{w}_k + \mathbf{M}_{\bar{w}}^{-1}(\mathbf{w}_k)\mathbf{M}(\mathbf{w}_k) \tag{4.88}$$

converges to a unique solution $\mathbf{w}^*$ of

$$\mathbf{x}_0 = \mathbf{M}(\mathbf{w}^*), \tag{4.89}$$

for which

$$\| \mathbf{w}_0 - \mathbf{w}^* \| \leqslant \frac{1}{h_0}(1 - \sqrt{1 - 2h_0})\eta_0. \tag{4.90}$$

Moreover, the rate of convergence is given by

$$\| \mathbf{w}^* - \mathbf{w}_k \| \leqslant (1/2)^{(k-1)}(2h_0)^{2(k-1)} \| \mathbf{w}_1 - \mathbf{w}_0 \|. \tag{4.91}$$

It is somewhat satisfying to know that under the conditions of Situation 1 the procedure will converge rapidly to $(\mathbf{v}^*, T^*)$. It is an advantage, from a computational point of view, to know that $y_k = 1$ under these conditions, since it removes the step of searching for a γ_k^* on each iteration.

Lemma 4.1

If the vector $\Delta\hat{\mathbf{y}}_k$ and the scalar $\Delta\hat{T}_k$ are such that the points $(\mathbf{v}_{k+1}(\gamma_k), T_{k+1}(\gamma_k))$ defined by Equations 4.77 and 4.78 are in the same subset $D_i \subset D$ as the point $(\mathbf{v}_k, T_k)$ (the current guess) for

---

[4] $|\tilde{\mathbf{M}}_\mathbf{w}| \equiv$ the determinant of the matrix $\tilde{\mathbf{M}}_\mathbf{w}$.

γ_k in $[0, \delta_k]$, $\delta_k > 0$ (not necessarily 1), then there exists a positive real number N_k such that the function $\|\mathbf{e}_{k+1}(y_k)\|$ is bounded on $[0, \delta_k]$ by the relation

$$\|\mathbf{e}_{k+1}(\gamma_k)\| \leq |1 - \gamma_k| \, \|\mathbf{e}_k\| + \gamma_k{}^2 N_k, \tag{4.92}$$

where $\mathbf{e}_k$ is defined by Equation 4.71.

Proof:

The function **M** can be considered a function of the single variable γ_k on the path g_k defined by the Equations 4.77 and 4.78, (namely, $\mathbf{M}(v_{k+1}(\gamma_k), T_{k+1}(\gamma_k))$. Using the chain rule, the fact that **M** is twice differentiable, and the mean-value theorem, and letting $\mathbf{M}(\gamma_k)$ denote $\mathbf{M}(v_{k+1}(\gamma_k), T_{k+1}(\gamma_k))$, we obtain

$$\mathbf{M}(\gamma_k) = \mathbf{M}(\gamma_k = 0) + \mathbf{M_v} \left.\frac{d\mathbf{v}_{k+1}}{d\gamma_k}\right|_{\gamma_k=0} \gamma_k + \mathbf{M}_T \left.\frac{dT_{k+1}}{d\gamma_k}\right|_{\gamma_k=0} \gamma_k$$

$$+ \tfrac{1}{2} \gamma_k{}^2 \mathbf{M}^{(2)}(\gamma), \tag{4.93}$$

where $\mathbf{M}^{(2)}(\gamma)$ is the second-derivative term evaluated at some $\gamma \in [0, \delta_k]$. But $\mathbf{e}_{k+1} = \mathbf{x}_0 - \mathbf{M}(\gamma_k)$, $\mathbf{e}_k = \mathbf{x}_0 - \mathbf{M}(0)$, $(dT_{k+1}/d\gamma_k)(\gamma_k) = \Delta\hat{T}_k$, and using Equation 4.77, we obtain $(d\mathbf{v}_{k+1}/d\gamma_k) = \Delta\hat{\mathbf{y}}_k$. Hence, we have

$$\mathbf{e}_{k+1}(\gamma_k) = \mathbf{e}_k - \mathbf{M_v}\Big|_{\gamma_k=0} \Delta\hat{\mathbf{y}}_k\, y_k - \mathbf{M}_T\Big|_{\gamma_k=0} \Delta\hat{T}_k\gamma_k - \tfrac{1}{2}\gamma_k{}^2\mathbf{M}^{(2)}(\gamma), \tag{4.94}$$

which becomes, in view of Equations 4.74, 4.75, and 4.76,

$$\mathbf{e}_{k+1} = (1 - \gamma_k)\mathbf{e}_k - \tfrac{1}{2}\gamma_k{}^2\mathbf{M}^{(2)}(\gamma), \tag{4.95}$$

or

$$N_k = \tfrac{1}{2} \, \|\mathbf{M}^{(2)}(\gamma)\|. \tag{4.96}$$

Q.E.D.

Lemma 4.2

If δ_k in Lemma 4.1 be less than 1 and positive such that if $\gamma_k \in [0, \delta_k)$, the point $(v_{k+1}(\gamma_k), T_{k+1}(\gamma_k))$ is in the same subset $D_l \subset D$ as $(\mathbf{v}_k, T_k)$ (the current guess); buth for γ_k in $(\delta_k, 1]$, the point $(\mathbf{v}_{k+1}(\gamma_k), T_{k+1}(\gamma_k))$ is another set $D_j \subset D$, and $(\mathbf{v}_{k+1}(\delta_k), T_{k+1}(\delta_k)) \in Q_T \cup Q_\eta$, then there exist two positive real numbers

ρ_k and Σ_k, such that the function $\| \mathbf{e}_{k+1}(\gamma_k) \|$ is bounded in $[\delta_k, 1]$ by the relation

$$\| \mathbf{e}_{k+1}(\gamma_k) \| \leq | 1 - \gamma_k | \, \| \mathbf{e}_k \| + \delta_k^2 N_k + (\gamma_k - \delta_k)\Sigma_k + (\gamma_k - \delta_k)^2 \rho_k, \tag{4.97}$$

where N_k was defined in Lemma 4. 1.

Proof:

Since $(\mathbf{v}_{k+1}(\delta_k), T_{k+1}(\delta_k)) \in \mathbf{Q}_T \cup \mathbf{Q}_\eta \cap D$, the following limits exist:

$$\lim_{\substack{\gamma_k \to \delta_k \\ \gamma_k \in [\delta_k, 1]}} \{\mathbf{M}_V(\mathbf{v}_{k+1}(\gamma_k), T_{k+1}(\gamma_k))\}, \tag{4.98}$$

$$\lim_{\substack{\gamma_k \to \delta_k \\ \gamma_k \in [\delta_k, 1]}} \{\mathbf{M}_T(\mathbf{v}_{k+1}(\gamma_k), T_{k+1}(\gamma_k))\}, \tag{4.99}$$

and

$$\lim_{\substack{\gamma_k \to \delta_k \\ \gamma_k \in [\delta_k, 1]}} \{\mathbf{M}^{(2)}(\mathbf{v}_{k+1}(\gamma_k), T_{k+1}(\gamma_k))\}, \tag{4.100}$$

and we can evaluate $\mathbf{e}_{k+1}(\gamma_k)$, y_k in $(\delta_k, 1]$ by using two Taylor-series expansions; that is,

$$\mathbf{M}(\delta_k) = \mathbf{M}(0) + \mathbf{M}_V \left. \frac{d\mathbf{v}_{k+1}}{d\gamma_k} \right|_{\gamma_k = 0} \delta_k + \mathbf{M}_T \Big|_{\gamma_k = 0} \Delta \hat{T}_k \delta_k + \tfrac{1}{2} \mathbf{M}^{(2)}(\gamma \in [0, \delta_k]) \delta_k^2, \tag{4.101}$$

and

$$\mathbf{M}(\gamma_k) = \mathbf{M}(\delta_k) + \mathbf{M}_V \left. \frac{d\mathbf{v}_{k=1}}{d\gamma_k} \right|_{\gamma_k = \delta_k} (\gamma_k - \delta_k) + \mathbf{M}_T \Big|_{\gamma_k = \delta_k} \Delta \hat{T}_k (\gamma_k - \delta_k) + \tfrac{1}{2} \mathbf{M}^{(2)}(\gamma \in [\delta_k, 1])(\gamma_k - \delta_k)^2. \tag{4.102}$$

But, we have

$$\left. \frac{d\mathbf{v}_{k+1}}{d\gamma_k} \right|_{\gamma_k = \delta_k} = ((1 - \beta^2)\Delta \hat{\mathbf{y}}_k - 2\mathbf{v}_k \beta^2 / \delta_k)/(1 + \beta^2)^{3/2}, \tag{4.103}$$

where

$$\beta = \frac{\gamma_k \| \Delta \hat{y}_k \|}{r}.$$

Hence, if we choose

$$\rho_k = \tfrac{1}{2}\mathbf{M}^{(2)}(\gamma \in [\delta_k, 1]) \tag{4.104}$$

and

$$\Sigma_k = \left\| \left. \mathbf{M_v} \frac{d\mathbf{v}_{k+1}}{d\gamma_k} \right|_{\gamma_k=\delta_k} - \left. \mathbf{M_v} \right|_{\gamma_k=0} \Delta \hat{\mathbf{y}}_k + \left(\left. \mathbf{M}_T \right|_{\gamma_k=\delta_k} - \left. \mathbf{M}_T \right|_{\gamma_k=0} \right) \Delta \hat{T} \right\|, \tag{4.105}$$

Relation 4.97 follows.

Equations 4.92 and 4.97 are illustrated in Figure 4.6.

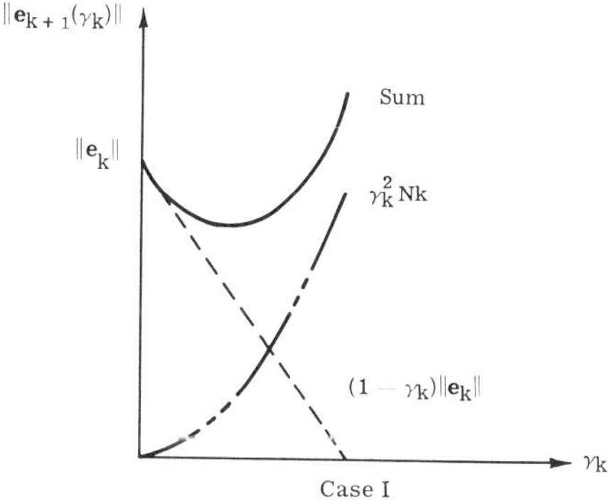

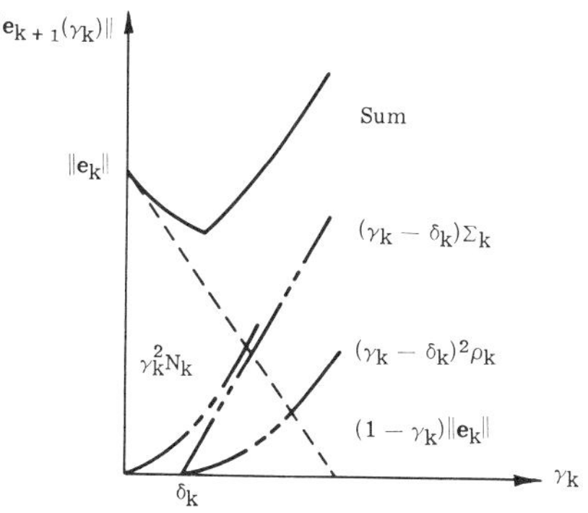

Figure 4.6 Illustration of the two types of bounds on $\|\mathbf{e}_{k+1}(\gamma_k)\|$.

Relation 4.97 does not bound $\| e_{k+1}(\gamma_k) \|$ if $(v_{k+1}(\delta_k), T_{k+1}(\delta_k)) \in Q_\mu$, because the limits of 4.98 and 4.100 "blow up." Suppose the Taylor expansion of Equation 4.102 is made about the point $\gamma_k = \delta_k + \epsilon$. Since **M** is continuous (Theorem 4.2), there exists a number $N(\epsilon)$ such that

$$\| \mathbf{M}(\delta_k + \epsilon) - \mathbf{M}(\delta_k) \| \leq N(\epsilon). \tag{4.106}$$

Hence, one can obtain a bound like that of Equation 4.97, when $(\mathbf{v}_{k+1}(\delta_k), T_{k+1}(\delta_k)) \in Q_\mu$ for γ_k in the interval $[\delta_k + \epsilon, 1]$ by using two Taylor expansions as in Lemma 4.2; that is,

$$\| e_{k+1}(\gamma_k \| \leq (1 - \gamma_k) \| e_k \| + \delta_k{}^2 N_k + N(\epsilon) + (\gamma_k - (\delta_k + \epsilon))\Sigma_k(\epsilon)$$
$$+ (\gamma_k - (\delta_k + \epsilon))^2{}_k(\epsilon). \tag{4.107}$$

Note, however, that $N(\epsilon)$ can be made arbitrarily small (that is, as ϵ decreases), but only at the expense of making $\Sigma_k(\epsilon)$ arbitrarily large.

The following theorem summarizes the results of this section.

Theorem 4.7

1. If $\delta_k \geq \frac{1}{2} \| e_k \| / N_k$, there is a $\gamma_k > 0$, such that

$$\| e_{k+1}(\gamma_k) \| < (1 - \gamma_k/2) \| e_k \|; \tag{4.108}$$

2. If $\delta_k < \frac{1}{2} \| e_k \| / N_k$ and $(v_{k+1}(\delta_k), T_{k+1}(\delta_k)) \notin Q_\mu$, there is a $\gamma_k \in [\delta_k, 1]$, such that

$$\| e_{k+1}(\gamma_k) \| < (1 - \delta_k/2) \| e_k \|. \tag{4.109}$$

3. If $\delta_k < \frac{1}{2} \| e_k \| / N_k$ and 2 does not hold, and if there is an ϵ such that $N(\epsilon) < \epsilon \| e_k \|$, there is a $\gamma_k \in [\delta_k, 1]$ such that,

$$\| e_{k+1}(\gamma_k) \| < \| e_k \|. \tag{4.110}$$

Proof:

(1) The minimum of the bound in Equation 4.92 occurs at $\gamma_k = \| e_k \| / 2N_k$, and this bound holds in all cases for $\gamma_k \in (0, \delta_k]$. Hence, if one chooses $\gamma_k = \| e_k \| / 2N_k$, Equation 4.108 follows from Equation 4.92.

(2) If $\delta_k < \| e_k \| / 2N_k$ and $(v_{k+1}(\delta_k), T_{k+1}(\delta_k)) \notin Q_\mu$ there are two cases depending on the slope of the right-hand side of Equation 4.97; either $\Sigma_k < \| e_k \|$, or $\Sigma_k \geq \| e_k \|$. Choose $\gamma_k = \delta_k$ if $\Sigma_k \geq \| e_k \|$, and $\gamma_k = \delta_k + (\| e_k \| - \Sigma_k)/2\rho_k$, if $\Sigma_k < \| e_k \|$. If $\delta_k < \| e_k \| / N_k$ and one chooses $\gamma_k = \delta_k$, the situation is precisely that of 1, and Equation 4.109 holds.

Using $\gamma_k = \delta_k + (\|e_k\| - \Sigma_k)/2\rho_k$, $\gamma_k \leq 1$, and $\delta_k < \|\mathbf{e}_k\|/N_k$ in Equation 4.97 and collecting terms, we obtain

$$\| \mathbf{e}_{k+1} \| < (1 - \delta_k/2) \| \mathbf{e}_k \| + (\| \mathbf{e}_k \| - \Sigma_k)(\Sigma_k - \| \mathbf{e}_k \|)/4\rho_k. \quad (4.111)$$

Hence, Equation 4.109 follows, since $\Sigma_k < \| \mathbf{e}_k \|$.

(3) Equation 4.110 follows directly from Equation 4.107 and the assumptions $\delta_k < \| \mathbf{e}_k \| / 2N_k$, $N(\epsilon) < \epsilon \| \mathbf{e}_k \|$ if $\gamma_k = \delta_k + \epsilon$.

It is assumed that δ_k is bounded away from zero. If the conditions of 1, 2, and 3 hold in Theorem 4.7, then the sequence $\| \mathbf{e}_k \|$ will decrease monotonically to zero. The rate of decrease will, however, be very slow if the N_k are large. Also, the procedure may experience difficulties in the vicinity of the set Q_μ if the condition on $N(\epsilon)$ in 3 does not hold.

4.6 Initial-Guess Procedures

Let T^*, $\mathbf{v}^*$, $\mathbf{x}_0$ denote the minimum time, the time-optimal terminal point on ∂S, and the initial state, respectively. A procedure for making an initial guess (v_1, T_1) based on the correlation between $e^{AT^*}x_0$, the state which the plant would achieve if a control $u_{(0,T^*]} = 0$ were applied, and $\mathbf{v}^*$ is proposed. From Equation 4.11, we have

$$\langle \mathbf{v}^*, e^{AT^*}\mathbf{x}_0 \rangle = \langle \mathbf{v}^*, \mathbf{v}^* \rangle + \int_0^T | \langle \mathbf{v}^*, e^{As}b \rangle | \, ds. \quad (4.112)$$

In other words,

$$\langle \mathbf{v}^*, e^{AT^*}\mathbf{x}_0 \rangle > r^2. \quad (4.113)$$

Suppose that $\tilde{T}^*$ was known fairly accurately. Then Equation 4.112 suggests that a good initial guess would be

$$T_1 = T^*, \quad (4.114)$$

$$v_1 = \frac{r e^{AT^*} \mathbf{x}_0}{\| e^{A\tilde{T}^*} \mathbf{x}_0 \|}. \quad (4.115)$$

Of course, T^* is not known for all $\mathbf{x}_0$ (if it were, the problem would be solved). However, it is possible to obtain bounds T_b on T^*.

<u>Upper Bounds on T^*.</u> An obvious upper bound on T^* is the time it takes for the trajectory to hit the target (in the case of asymptotically stable systems) in the absence of any control. The quality of such a bound will be better for large r, for eigenvalues with large negative real parts, and if the gain is low.

<u>Lower Bounds on T^*</u> (The method of problem separation). Suppose

that the matrix $\mathbf{A}$ is composed of blocks $\mathbf{A}_i$ of lesser-dimensional matrices on the diagonal with zeros elsewhere; that is, suppose that

$$\mathbf{A} = \begin{bmatrix} \mathbf{A}_1 & & & & 0 \\ & \mathbf{A}_2 & & & \\ & & \cdot & & \\ & & & \cdot & \\ & & & & \cdot \\ 0 & & & & \mathbf{A}_l \end{bmatrix} \tag{4.116}$$

We consider that such a matrix is "separable" into the blocks A_i, $i = 1, 2, \ldots, l$. With such a plant and Problem Z_1, we can associate l plants and l problems Z_{1i}, where the ith plant is

$$\dot{\mathbf{x}}_i = \mathbf{A}_i\mathbf{x}_i + \mathbf{b}_i u, \qquad |u| \leq 1, \qquad \mathbf{x}_i(0) = \mathbf{x}_{0i},$$

$$i = 1, 2, \ldots, l, \tag{4.117}$$

$$S = \{\mathbf{x}_i : \langle \mathbf{x}_i, \mathbf{x}_i \rangle - r^2 \leq 0\}. \tag{4.118}$$

Since both of these are problems of lower dimension, they are easier to solve for T_i^* than the original. For example, if the kth separated problem has dimension 1, we have

$$T_k^* = \frac{1}{|\lambda_k|} \log \left\{ \frac{1 + |x_{0k}|}{1 + r} \right\}. \tag{4.119}$$

If each such separated problem is solved for a T_i^*, then, clearly, if

$$T_b = \max_{i=1,2,\ldots,l} \{T_i^*\}, \tag{4.120}$$

we have

$$T_b \leq T^*, \tag{4.121}$$

where T^* is the minimum time for the original problem. This statement can be proved by contradiction. Suppose that $T_b = T_m^*$ in Equation 4.120 and suppose that $T^* < T_m^*$. Then, there exists a $u_{(0,T^*]}$ which steers the state in the separated problem m from the initial condition $\mathbf{x}_{0m}$ to a point in S_m in less than the optimal time T_m^*. This is a contradiction. The proposed initial-guess procedure is as follows.

1. Find T_b using the method of problem separation,

2. Find $e^{\mathbf{A}T_b}\mathbf{x}_0$;

3. Choose $T_1 = T_b$ and

$$\mathbf{v}_1 = \frac{r e^{\mathbf{A}T_b}\mathbf{x}_0}{\|e^{\mathbf{A}T_b}\mathbf{x}_0\|}.$$

4.7 Generalization to the Multi-input Case

The procedure generalizes easily to the multi-input case. Suppose that the control $\mathbf{u}$ is an m-vector; that is,

$$\dot{\mathbf{x}}(t) = \mathbf{A}\mathbf{x}(t) + \mathbf{B}\mathbf{u}(t), \tag{4.122}$$

where $\mathbf{B}$ is an $n \times m$ matrix. Equation 4.122 can also be written

$$\dot{\mathbf{x}}(t) = \mathbf{A}\mathbf{x}(t) + \sum_{i=1}^{m} \mathbf{b}_i u_i(t), \tag{4.123}$$

where $\mathbf{b}_i$ is the ith column of $\mathbf{B}$. The Hamiltonian is given by

$$H = \mathbf{p}_0 + \langle \mathbf{p}(t), \mathbf{A}\mathbf{x}(t)\rangle + \sum_{i=1}^{m} \langle \mathbf{p}(t), \mathbf{b}_i\rangle u_i(t), \tag{4.124}$$

and

$$u_i^*(t) = -\text{sgn}\{\langle \mathbf{p}^*(t), \mathbf{b}_i\rangle\}, \qquad i = 1, 2, \ldots, m. \tag{4.125}$$

The function $\mathbf{M}$ becomes

$$\mathbf{M}(\mathbf{v}, T) = e^{-\mathbf{A}T}\mathbf{v} + e^{-\mathbf{A}T} \sum_{i=1}^{m} e^{\mathbf{A}s}\mathbf{b}_i \ \text{sgn}\{\langle \mathbf{v}, \mathbf{Q}e^{\mathbf{A}s}\mathbf{b}_i\rangle\}\, ds. \tag{4.126}$$

Theorem 4.1 remains the same, except that Relation 4.23 becomes

$$-\langle \mathbf{v}, \mathbf{QAv}\rangle + \sum_{i=1}^{m} |\langle \mathbf{v}, \mathbf{Q}\mathbf{b}_i\rangle| \geq 0. \tag{4.127}$$

Also, $\mathbf{M}(\mathbf{v}, T)$ of Equation 4.126 is continuous on D. The proof of this theorem parallels that of Theorem 4.2.

Clearly, the set of matrices $\mathbf{A}$ for which

$$-\langle \mathbf{v}, \mathbf{QAv}\rangle + \sum_{i=1}^{m} |\langle \mathbf{v}, \mathbf{Q}\mathbf{b}_i\rangle| > 0, \qquad m > 1, \tag{4.128}$$

is larger than that for which Relation 4.23 holds. For example, if $\mathbf{B}$ has rank n, Relation 4.128 can be satisfied everywhere on ∂S even if the eigenvalues of $\mathbf{A}$ are all in the right half-plane. Proof that $\mathbf{M}$ is one-to-one on D follows that of Theorem 4.3.

Definition 4.3(a) (See Definition 4.3.)

Let $\eta(\mathbf{v})$ be a function from D into R_m, such that

$$\eta_i(\mathbf{v}) = \begin{cases} \text{sgn}\{\langle \mathbf{v}, \mathbf{Q}\mathbf{b}_i\rangle\}, & \langle \mathbf{v}, \mathbf{Q}\mathbf{b}_i\rangle \neq 0, \\ \text{sgn}\{\langle \mathbf{v}, \mathbf{Q}e^{\mathbf{A}(\tau=0^+)}\mathbf{b}_i\rangle\}, & \langle \mathbf{v}, \mathbf{Q}\mathbf{b}_i\rangle = 0, \end{cases} \qquad i = 1, 2, \ldots, m. \tag{4.129}$$

Definition 4.4(a) (Q_η)

Let $Q_\eta \subset D$ be a set of ordered pairs $(\mathbf{v}, T)$, such that, for some b_l in $\{b_i\}_1^m$,

$$\langle v, \mathbf{Q}b_l \rangle = 0. \tag{4.130}$$

From Definition 4.4(a), Q_η is the union of m hyperplanes.

Definition 4.5(a) $(\mu(\mathbf{v}, T))$

Let $\mu(\mathbf{v}, T)$ be a function from D into R_m whose elements $\mu_i(\mathbf{v}, T)$ are integers, such that the sequence $u_{i(0,T]}$ corresponding to the ordered pair $(\mathbf{v}, T)$ makes exactly $\mu_i(\mathbf{v}, T)$ switchings in the open interval $(0,T)$.

Definition 4.6(a) $(\tau_{ij}(\mathbf{v}), i = 1, 2, \ldots, m, j = 1, 2, \ldots, \mu_i(\mathbf{v}, T)$

The switch times τ_{ij}, $i = 1, 2, \ldots, m, j = 1, 2, \ldots, \mu_i(\mathbf{v}, T)$ are a set of ordered real numbers such that

$$1. \quad \tau_{ij}(\mathbf{v}) \in (0, T), \tag{4.131}$$

$$2. \quad \langle \mathbf{v}, \mathbf{Q}e^{\mathbf{A}\tau_{ij}(\mathbf{v})}\mathbf{b} \rangle = 0, \tag{4.132}$$

$$3. \quad \operatorname{sgn}\{\langle \mathbf{v}, \mathbf{Q}e^{\mathbf{A}\tau_{ij}^{+}(\mathbf{v})}\mathbf{b} \rangle\} = -\operatorname{sgn}\{\langle \mathbf{v}, Qe^{A\tau_{ij}^{-}(\mathbf{v})}\mathbf{b} \rangle\} \tag{4.133}$$

$$4. \quad \tau_{ij}(\mathbf{v}) < \tau_{ij+1}(\mathbf{v}) \quad \text{for all j}. \tag{4.134}$$

Definition 4.7(a) (Q_T)

Let $Q_T \subset D$ be a set of ordered pairs $(\mathbf{v}, T)$, such that $(\mathbf{v}, T)$ satisfies Equations 4.132 and 4.133.

Definition 4.8(a) (Q_μ)

Let $Q_\mu \subset D$ be a set of ordered pairs $(\mathbf{v}, T)$, such that for one or more times τ in $(0, T)$ and one or more $\mathbf{b}_i$'s in $\{\mathbf{b}_i\}_{i=1,2,\ldots,m}$, we have the equations

$$\langle \mathbf{v}, \mathbf{Q}e^{\mathbf{A}\tau}\mathbf{b}_i \rangle = 0, \tag{4.135}$$

$$\langle \mathbf{v}, \mathbf{Q}\mathbf{A}e^{\mathbf{A}\tau}\mathbf{b}_i \rangle = 0. \tag{4.136}$$

Definition 4.9(a) (Q)

Let

$$Q = Q_T \cup Q_\eta \cup Q_\mu. \tag{4.137}$$

The set Q has dimension $n - 1$ as before. There are, in effect, m times as many surfaces in $\mathbf{Q}$ as there were in the scalar u case. The sets D_i are similarly defined. Also, $\mathbf{M}$ is piecewise differentiable to all orders, as before, on the open sets D_i.

The matrix $\mathbf{G}$ becomes

$$\mathbf{G} = \left[re^{-\mathbf{A}T} \left\{ \mathbf{I} + \sum_{i=1}^{m} \mathbf{E}_i \mathbf{L}^{-1} \right\} \mathbf{h}_\theta(\theta) \,\Big|\, \mathbf{M}_T(\mathbf{v}, T) \right], \tag{4.138}$$

where

$$\mathbf{E}_i = \sum_{j=1}^{m} \frac{2}{|\langle \mathbf{v}, \mathbf{QA}e^{\mathbf{A}\tau_{ij}(\mathbf{v})}\mathbf{b}_i \rangle|} e^{\mathbf{A}\tau_{ij}(\mathbf{v})}\mathbf{b}_i\rangle \langle \mathbf{Q}e^{\mathbf{A}\tau_{ij}(\mathbf{v})}\mathbf{b}_i,$$

$$i = 1, 2, \ldots, m. \tag{4.139}$$

Each matrix $\mathbf{E}_1$ is positive semidefinite as before, and $\mathbf{G}$ will have rank n (as before) if relation 4.127 holds with a strict inequality.

Sections 4.3 and 4.4 apply unchanged to this case. One can conclude that for $m > 1$, $D = \partial S \times R_1^+$ for a larger set of matrices $\mathbf{A}$ but at the expense of increasing the size of the set Q. The initial-guess procedure of Section 4.5 can also be applied equally well to the multi-input case.

That this iterative procedure adapts to the multi-input case is a point in its favor. Also, note from Equations 4.26 and 4.139 that the evaluation of $\mathbf{M}(\mathbf{v}, T)$ and the evaluation of $G(\mathbf{v}, T)$ increases only linearly with the number of controllers m.

4.8 The Time-Varying Case

In this section, it is assumed that $\mathbf{Q} = \mathbf{I}$ and $m = 1$; that is,

$$\dot{\mathbf{x}}(t) = \mathbf{A}(t)\mathbf{x}(t) + \mathbf{b}(t)u(t), \qquad \mathbf{x}(t_0) = \mathbf{x}_0, \tag{4.140}$$

and

$$S = (\mathbf{x}: \|\mathbf{x}\| - r \leqslant 0), \qquad r > 0. \tag{4.141}$$

These assumptions are made to simplify the presentation and are not necessary. The function $\mathbf{M}$ becomes (see Appendix B)

$$\mathbf{M}(\mathbf{v}, T) = \mathbf{X}(t_0)\mathbf{X}^{-1}(T)(\mathbf{v} + \int_{t_0}^{T} \mathbf{X}(T)\mathbf{X}^{-1}(s)\mathbf{b}(s)$$

$$\operatorname{sgn}(\langle \mathbf{v}, \mathbf{X}(T)\mathbf{X}^{-1}(s)\mathbf{b}(s)\rangle)\, ds), \tag{4.142}$$

where $\mathbf{X}(t)$ satisfies the matrix differential equation

$$\dot{\mathbf{X}}(t) = \mathbf{A}(t)\mathbf{X}(t), \qquad \mathbf{X}(0) = \mathbf{I}. \tag{4.143}$$

Equation 4.142 has a slightly different form than Equation 4.11 in that reverse time has not been used. More explicitly, if it is assumed that $\mathbf{M}(\mathbf{v}, T)$ is the starting point for an extremal trajectory that terminates on $\mathbf{v} \in \partial S$ at $t = T$, then we have

$$\mathbf{v} = \mathbf{X}(T)\mathbf{X}^{-1}(t_0)\mathbf{M}(\mathbf{v}, T) - \int_{t_0}^{T} \mathbf{X}(T)\mathbf{X}^{-1}(s)\mathbf{b}(s)$$

$$\operatorname{sgn}(\langle \mathbf{p}(s), \mathbf{b}(s)\rangle)\, ds, \tag{4.144}$$

where

$$\dot{\mathbf{p}}(t) = -\mathbf{A}'(t)\mathbf{p}(t).$$

Hence, we have

$$\mathbf{p}(t) = (\phi'(t, t_0))^{-1}\mathbf{p}(0),$$

or

$$\mathbf{p}(0) = \phi'(T, t_0)\mathbf{p}(T) = \mathbf{X}'^{-1}(t_0)\mathbf{X}'(T)\mathbf{v},$$

and

$$\mathbf{p}(t) = \mathbf{X}'^{-1}(t)\mathbf{X}'(T)\mathbf{v}. \tag{4.145}$$

Equation 4.142 is obtained after substituting Equation 4.145 into Equation 4.144 and multiplying the result by $\mathbf{X}(t_0)\mathbf{X}^{-1}(T)$.

A theorem similar to Theorem 4.2 is easily derived, if one assumes that all of the eigenvalues of $\mathbf{A}(t)$ have negative real parts for all time. Relation 4.16 becomes

$$\langle \mathbf{v}_1, \mathbf{Q}\mathbf{X}(T)\mathbf{X}^{-1}(t_0)\mathbf{v}_2\rangle \leq \langle \mathbf{v}_1, \mathbf{Q}\mathbf{v}_1\rangle, \tag{4.146}$$

and the portion of that proof up to and including Equation 4.22 holds with $e^{\mathbf{A}(\tau_i - s)}$ replace by $\mathbf{X}(\tau_i)\mathbf{X}^{-1}(s)$, $i = 1, 2 \ldots$. Now, p_0 must be nonnegative. Hence, evaluating the Hamiltonian at $t = T$, we find that NC9 is satisfied if and only if

$$-\langle \mathbf{v}, \mathbf{Q}\mathbf{A}(T)\mathbf{v}\rangle + |\langle \mathbf{v}, \mathbf{Q}\mathbf{b}(T)\rangle| \geq 0. \tag{4.147}$$

Again, this will hold if $\langle \mathbf{x}, \mathbf{Q}\mathbf{x}\rangle$ is a Lyapunov function for the system $\dot{\mathbf{x}}(t) = \mathbf{A}(t)\mathbf{x}(t)$. (Recall that if $\langle \mathbf{x}\mathbf{Q}\mathbf{x}\rangle$ is a Lyapunov function, then $\langle \mathbf{x}(t), \mathbf{Q}\mathbf{A}(t)\mathbf{x}(t)\rangle < 0$, for all t and all $\mathbf{x} \in R_n$.) The remainder of the proof procedes in precisely the same manner as the proof in Theorem 4.2.

Theorems 4.2 and 4.3 do not require the plant to be time-invariant and hence extend without difficulty to the time-varying case. The function $\eta(\mathbf{v})$ and the set Q_η are defined exactly as in Definitions 4.3 and 4.4. The function $\mu(\mathbf{v}, T)$ is defined as in Definition 4.5, except that it becomes a function of t_0 [that is, $\mu(\mathbf{v}, T, t_0)$]. The switch times

τ_j become functions of $\mathbf{v}$, t_0, and T, and are defined as in Definition 4.6 by replacing $e^{\mathbf{A}\tau_j(\mathbf{v})}\mathbf{b}$ by $\mathbf{X}(T)\mathbf{X}^{-1}(t_j(\mathbf{v}, T, t_0))\mathbf{b}$, where the symbol t_j is used since reverse time is not. [That is, $\tau_j(\mathbf{v}, T, t_0) = T + t_0 - t_j(\mathbf{v}, T, t_0)$]. The set Q_T is redefined as follows. Let $Q_T \subset D$ be the set of ordered pairs $(\mathbf{v}, T)$, such that $(\mathbf{v}, t_0)$ satisfies Equations 4.33 and 4.44, and Q_μ is defined as in Definition 4.8, with $e^{\mathbf{A}\tau}\mathbf{b}$ and $\mathbf{A}e^{\mathbf{A}\tau}\mathbf{b}$ replaced by $\mathbf{X}(T)\mathbf{X}^{-1}(t)\mathbf{b}(t)$ and $\mathbf{X}(T)\mathbf{X}^{-1}(t)\,(-\mathbf{A}(t)\mathbf{b}(t) - \dot{\mathbf{b}}(t))$, respectively ($\mathbf{b}(t)$ is assumed to be differentiable). The sets Q_η and Q_T have dimension $n - 1$, and the set Q_μ will have dimension $n - 1$ if the vectors $\mathbf{X}(T)\mathbf{X}^{-1}(t)\mathbf{b}(t)$ and $\mathbf{X}(T)\mathbf{X}^{-1}(t)\,(-A(t)\mathbf{b}(t) + \dot{\mathbf{b}}(t))$ are linearly independent for all t (see the discussion following Definition 4.8).

Theorem 4.4 also follows easily with appropriate changes, and Equation 4.52 becomes

$$\frac{\partial \tau_j}{\partial v_i} = -\frac{\langle \mathbf{i}, \mathbf{Q}\mathbf{X}(T)\mathbf{X}^{-1}(t_j)\mathbf{b}(t_j)\rangle}{\langle \mathbf{v}, \mathbf{Q}\mathbf{X}(T)\mathbf{X}^{-1}(t_j)(-\mathbf{A}(t_j)\mathbf{b}(t_j) + \dot{\mathbf{b}}(t_j))\rangle}. \tag{4.148}$$

In Theorem 4.5, Equation 4.65 becomes

$$\mathbf{E} = \sum_{j=1}^{\mu} \frac{2\mathbf{X}(T)\mathbf{X}^{-1}(t_j)\mathbf{b}(t_j)\rangle\langle\mathbf{Q}\mathbf{X}(T)\mathbf{X}^{-1}(t_j)\mathbf{b}(t_j)}{|\langle \mathbf{v}, \mathbf{Q}\mathbf{X}(T)\mathbf{X}^{-1}(t_j)(-\mathbf{A}(t_j)\mathbf{b}(t_j) + \dot{\mathbf{b}}(t_j))\rangle|} \tag{4.149}$$

Since $(\mathbf{X}(T)\mathbf{X}^{-1}(t_0))'\mathbf{Q}\mathbf{v}$ is orthogonal to the first $n - 1$ columns of $\mathbf{G}$, the remainder of the theorem follows if

$$\langle \mathbf{v}, \mathbf{Q}\mathbf{X}(T)\mathbf{X}^{-1}(t_0)\mathbf{M}_T(\mathbf{v}, (T)\rangle \neq 0. \tag{4.150}$$

But,

$$\begin{aligned}\mathbf{M}_T = &-\mathbf{X}(t_0)\mathbf{X}^{-1}(T)\mathbf{A}(T)\\ &\times (\mathbf{v} + \int_{t_0}^{T} \mathbf{X}(T)\mathbf{X}^{-1}(s)\mathbf{b}(s)\ \mathrm{sgn}(\langle \mathbf{v}, \mathbf{Q}\mathbf{X}(T)\mathbf{X}^{-1}(s)\mathbf{b}(s)\rangle)ds\\ &+ \mathbf{X}(t_0)\mathbf{X}^{-1}(T)\mathbf{b}(T)\ \mathrm{sgn}(\langle \mathbf{v}, \mathbf{Q}\mathbf{b}(T)\rangle)).\end{aligned} \tag{4.151}$$

Hence, we have

$$\begin{aligned}\langle \mathbf{v}, \mathbf{Q}\mathbf{X}(T)\mathbf{X}^{-1}(t_0)\mathbf{M}_T(\mathbf{v}, T)\rangle = &-\langle \mathbf{v}, \mathbf{Q}\mathbf{A}(T)\mathbf{v}\rangle + |\langle \mathbf{v}, \mathbf{Q}\mathbf{b}(T)\rangle|\\ &+ \int_{t_0}^{T} |\langle \mathbf{v}, \mathbf{Q}\mathbf{X}(T)\mathbf{X}(s)\mathbf{b}(s)\rangle|\, ds,\end{aligned} \tag{4.152}$$

which is clearly positive if $\langle \mathbf{x}, \mathbf{Q}\mathbf{x}\rangle$ is a Lyapunov function for the system $\dot{\mathbf{x}}(t) = \mathbf{A}(t)\mathbf{x}(t)$. The fact that $\mathbf{A}(t)$ and $\mathbf{b}(t)$ are time-varying does not affect the iterative procedure of Section 4.4, the convergence arguments of Section 4.5, or the initial-guess procedures of Section 4.6. Hence, it may be concluded that the technique will be equally useful in the time-varying case.

4.9 Singular Time-Optimal Control

It has been assumed so far that the problem was normal. This permitted us to define **M** using the control law which arose out of the minimization of the Hamiltonian. The fact that the minimization of the Hamiltonian does not yield an optimal-control law does not mean that one does not exist. A pertinent question then is: Does an **M** exist when the problem is singular? The following example illustrates a class of systems for which **M** is not a function on D when the problem is singular.

Example 4.4 Let the plant in Problem Z_1 be

$$\begin{bmatrix} \dot{x}_1 \\ \dot{x}_2 \end{bmatrix} = \begin{bmatrix} \lambda & 0 \\ 0 & \lambda \end{bmatrix} + \begin{bmatrix} b_1 \\ b_2 \end{bmatrix} u, \qquad |u| \leq 1, \tag{4.153}$$

where

$$\lambda < 0, \qquad \text{and} \qquad S = \{\mathbf{x}: \langle \mathbf{x}, \mathbf{x} \rangle - r^2 \leq 0\}, \qquad r > 0.$$

The controllability matrix for this plant is

$$C = \begin{bmatrix} b_1 & \lambda b_1 \\ b_2 & \lambda b_2 \end{bmatrix}, \tag{4.154}$$

which is singular for all λ and all $\mathbf{b}$. Hence, the problem is singular. The trajectories for $u = \pm 1 \equiv \Delta$ satisfy

$$\frac{dx_2}{dx_1} = \frac{\lambda x_2 + b_2 \Delta}{\lambda x_1 + b_1 \Delta}. \tag{4.155}$$

These are straight lines passing through the points $(-b_1\Delta/\lambda, -b_2\Delta/\lambda)$. If the target set has radius $r = 0$, then an optimal control exists only if the initial state $\mathbf{x}_0$ lies between the points $(-b_1/\lambda, -b_2/\lambda)$ and $(b_1/\lambda, b_2/\lambda)$ on the line segment joining these two points. However, if the target set has an arbitrarily small radius, there exists an optimal control to S for every state in the state space. The solution of this problem for $r > 0$ is given in Figure 4.7. From Figure 4.7 it is seen that many optimal trajectories terminate on the points

$$\mathbf{v} = \frac{\pm r}{\|\mathbf{b}\|} \begin{bmatrix} b_2 \\ -b_1 \end{bmatrix}.$$

It follows that **M** is not a function on D in this case.

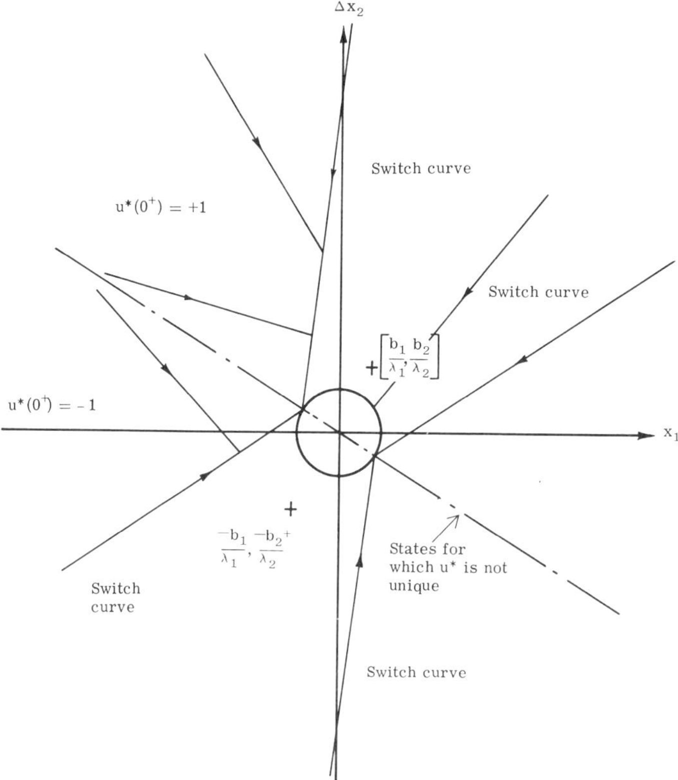

Figure 4.7 Singular time-optimal control.

Example 4.5. Let the plant in Problem Z_1 be

$$\begin{bmatrix} \dot{x}_1 \\ \dot{x}_2 \end{bmatrix} = \begin{bmatrix} \lambda_1 & 0 \\ 0 & \lambda_2 \end{bmatrix} + \begin{bmatrix} b_1 \\ b_2 \end{bmatrix} u, \qquad |u| \leqslant 1, \tag{4.156}$$

where

$$\lambda_1 = \lambda \text{ of Example 5.3,}$$

and

$$\lambda_2 = \lambda - \epsilon,$$

Choose ϵ small enough so that the trajectories can be drawn (for illustration purposes) as straight lines. It is interesting to compare the solution of this problem (where **M** is one-to-one on D) in Figure 4.8 with that in Figure 4.7.

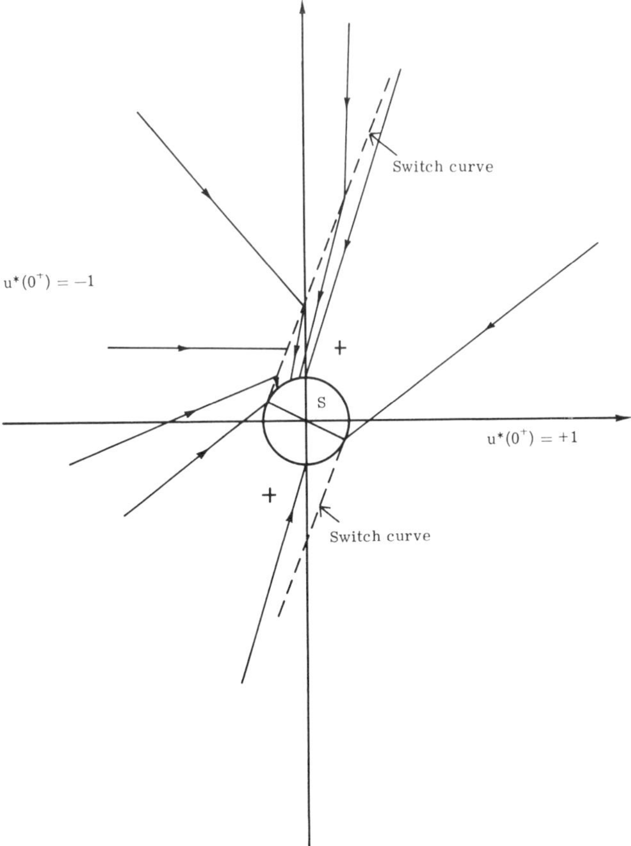

Figure 4.8 "Almost singular" time-optimal trajectories.

4.10 The Unmodified Time-Optimal Problem

In the unmodified time-optimal regulator problem, the target set is the origin. It is interesting to see how the fact that **M** is not one-to-one in this case [22] affects the iterative procedure. The form of **M** and its derivatives can be easily deduced for this problem. Let S_p be a unit hypersphere in R_n; that is,

$$S_p = \{\boldsymbol{\pi}; \pi \in R_n, \langle \boldsymbol{\pi}, \pi \rangle = 1\}, \tag{4.157}$$

where $\boldsymbol{\pi}$ is the boundary condition on the costate differential equation. Thus,

$$\dot{\mathbf{p}}(\tau) = \mathbf{A}'\mathbf{p}(\tau), \qquad \mathbf{p}(0) = \boldsymbol{\pi}. \tag{4.158}$$

Let $\tilde{\mathbf{M}}$ be a function whose domain is all of $R_n \times R_1^+/\phi$ (see Definition 4.1) defined as

$$\tilde{\mathbf{M}}(\pi, T) = e^{-\mathbf{A}T} \int_0^T e^{\mathbf{A}s}\mathbf{b}\ \mathrm{sgn}\{\langle \pi, e^{\mathbf{A}s}\mathbf{b}\rangle\}\, ds. \tag{4.159}$$

In other words,

$$\tilde{\mathbf{M}}(\pi, T) = \mathbf{M}_2(\pi, T). \tag{4.160}$$

If we proceed in a fashion similar to that of Section 4.3, we have

$$\tilde{\mathbf{G}} = [e^{-\mathbf{A}T}\mathbf{E}\mathbf{h}_\theta(\theta)\tilde{\mathbf{M}}_T] \tag{4.161}$$

for $(\mathbf{v}, T) \notin Q$. However, there is no longer any guarantee that $\mathbf{G}$ will have rank n. In fact, if μ is less than n, the rank of $\mathbf{G}$ cannot be n. This follows from the fact that $\mathbf{E}$ is the sum of μ dyad matrices (see Equation 4.63). Since a dyad matrix has rank 1, a sum of less than n of them cannot have rank n. This means immediately that this approach cannot be used for plants with only real eigenvalues, because in such plants μ can be at most $n - 1$ [38]. For plants with complex eigenvalues, there is a possibility that $\mathbf{G}$ will have rank n when $\mu \geq n$; however, no guarantees have been found that this will be the case. (The reasoning concerning the rank of the equivalent to $\mathbf{G}$ in Reference 22 is incorrect.)

4.11 Summary

The time-optimal control problem Z_1 has been treated in some detail in this chapter. It has been shown that $\mathbf{M}(\mathbf{v}, T)$ is a one-to-one continuous function which is piecewise continuously differentiable to all orders on D. The set D was carefully defined, and sufficient conditions for D to be such that the proposed iterative procedure will be practical (that is, $D = \partial S \times R_1^+$) have been presented. A set $Q \subset D$ on which the derivatives of $\mathbf{M}(\mathbf{v}, T)$ are discontinuous has been found, and it was shown that this "bad set" has dimension $n - 1$, where n is the dimension of D. An iterative procedure was proposed which was based on the fact that the matrix $\mathbf{G}(\mathbf{v}, T)$ is nonsingular wherever it exists. (which is almost everywhere on D), and conditions guaranteeing monotone or almost monotone convergence were given. A method of initializing the iterative procedure was also discussed. All results were extended to include both the multicontroller and the time-varying plant. Finally, the modified problem was compared with the unmodified one with respect to the existence of $\mathbf{G}^{-1}(\mathbf{v}, T)$, and it was shown that the inverse of $\mathbf{G}$ may not exist in the latter case.

5. The Time-Optimal Problem (Ω_b)

5.1 Introduction

Problem Z_2 is investigated in this chapter (see Definition 1.12). The format of Chapter 4 is followed, and that chapter is referenced wherever possible. Throughout the discussion, an attempt is made to comment on the similarities and differences between Problems Z_1 and Z_2, insofar as the iterative procedure is concerned. The main features of Problem Z_2, in comparison to those of Z_1, are that the set Q can have lower dimension in Problem Z_2, and that the matrix $\mathbf{G}(\mathbf{v}, T)$ and the function $\mathbf{M}(\mathbf{v}, T)$ are more difficult to evaluate in Problem Z_2.

5.2 The Function M

Obviously, as in Problem Z_1, the time-optimal terminal state $\mathbf{x}^*(T) \in \partial S$, and hence the transversality condition NC10 of Theorem 1.2 applies, that is,

$$\mathbf{x}^*(T^*) = \mathbf{v}^*, \qquad \mathbf{v}^* \in \partial S, \tag{5.1}$$

and

$$\mathbf{p}^*(T^*) = \alpha \mathbf{v}^*, \qquad \alpha > 0. \tag{5.2}$$

The Hamiltonian is (see Definition 1.12 and Equation 1.17)

$$H = p_0 + \langle \mathbf{p}(t), \mathbf{A}\mathbf{x}(t) \rangle + \langle \mathbf{p}(t), \mathbf{B}\mathbf{u}(t) \rangle, \tag{5.3}$$

whence

$$\dot{\mathbf{p}}^*(t) = -\mathbf{A}'\mathbf{p}^*(t), \qquad \mathbf{p}^*(T^*) = \alpha \mathbf{v}^*, \tag{5.4}$$

$$\dot{\mathbf{x}}^*(t) = \mathbf{A}\mathbf{x}^*(t) + \mathbf{B}\mathbf{u}^*(t), \qquad \mathbf{x}^*(0) = \mathbf{x}_0. \tag{5.5}$$

From NC7 of Theorem 1.2, and from the fact that $\mathbf{u} \in \Omega_b$ (that is, $\|u\|_R \leq 1$), we have

$$\mathbf{u}^*(t) = \frac{\mathbf{R}^{-1}\mathbf{B}'\mathbf{p}^*(t)}{\|\mathbf{B}'\mathbf{p}^*(t)\|_{\mathbf{R}}}. \tag{5.6}$$

Conditions on the normality of Problem Z_2 are given in Theorem C.2 of Appendix C. Since normality is assumed, $\mathbf{u}^*(t)$ in Equation 5.6 is well defined for all initial conditions $\mathbf{v}^* \neq \mathbf{0}$, except at the finite number of times (possibly zero) for which

$$\mathbf{B}'\mathbf{p}^*(t) = \mathbf{0}. \tag{5.7}$$

Since Equations 5.4, 5.5, and 5.6 are independent of $|\alpha|$ and p_0, these parameters are not needed for defining $\mathbf{M}$. Choose a $\mathbf{v} \neq \mathbf{0}$ in R_n and a $T \in R_1^+$, and let

$$\tau = T - t; \tag{5.8}$$

Then, we have

$$\dot{\mathbf{p}}(\tau) = \mathbf{A}'\mathbf{p}(\tau), \qquad \mathbf{p}(\tau = 0) = \mathbf{v}. \tag{5.9}$$

Hence, we obtain

$$\mathbf{p}(\tau) = e^{\mathbf{A}'\tau}\mathbf{v}. \tag{5.10}$$

Let

$$\mathbf{F}'(\tau) \equiv \mathbf{R}^{-1}\mathbf{B}'e^{\mathbf{A}'\tau}. \tag{5.11}$$

Then, the initial-value problem defining $\mathbf{M}$ is

$$\dot{\mathbf{x}}(\tau) = -\mathbf{A}\mathbf{x}(\tau) - \mathbf{B}\mathbf{u}(\tau), \tag{5.12}$$

$$\mathbf{u}(\tau) = -\frac{\mathbf{F}'(\tau)\mathbf{v}}{\|\mathbf{F}'(\tau)\mathbf{v}\|_{\mathbf{R}}}. \tag{5.13}$$

Definition 5.1

Let $\mathbf{M}$ be a function whose domain is $\partial S \times R_1^+$ (see Definition 4.1) defined as

$$\mathbf{M}(\mathbf{v}, T) = e^{-\mathbf{A}T}\left\{\mathbf{v} + \int_0^T \frac{\mathbf{F}(s)\mathbf{F}'(s)\mathbf{v}\,ds}{\|\mathbf{F}'(s)\mathbf{v}\|_{\mathbf{R}}}\right\}; \tag{5.14}$$

that is, $\mathbf{M}(\mathbf{v}, T)$ is a solution, at time $\tau = T$, of Equations 5.12 and 5.13.

5.3 Definitions and Preliminary Theorems

From Equation 5.13 and the normality, it follows that given $\mathbf{v} \neq \mathbf{0}$, the control $\mathbf{u}_{(0,T]}$ is a piecewise continuous function from (0, T] into R_m, such that $\|\mathbf{u}(\tau)\|_{\mathbf{R}} = 1$ almost everywhere in (0, T]. The discontinuities, if any, occur at times $\tau_j(\mathbf{v})$, such that

$$\mathbf{B}'e^{\mathbf{A}\tau_j(\mathbf{v})}\mathbf{v} = 0. \tag{5.15}$$

At such times, the vector $\mathbf{u}(\tau)$ reverses its direction, that is,

$$\lim_{\epsilon \to 0} \{\mathbf{u}(\tau_j + \epsilon)\} = -\lim_{\epsilon \to 0} \{\mathbf{u}(\tau_j - \epsilon)\}. \tag{5.16}$$

This situation is similar to that in Problem Z_1, where the m inputs all switch at the same time.

The set D is defined exactly as in Definition 4.2. A theorem like that of Theorem 4.1 follows if Equation 4.19 is replaced by Equation 5.13, and if Equation 4.23 is replaced by

$$-\langle \mathbf{v}, \mathbf{A}\mathbf{v}\rangle + \|\mathbf{B}\mathbf{v}\|_{\mathbf{R}} > 0. \tag{5.17}$$

Furthermore, $\mathbf{M}(\mathbf{v}, T)$ is seen to be continuous in the same manner as it is in Theorem 4.2, and Theorem 4.3 applies here.

Definition 5.2 (Q)

Let $Q \subset D$ be the set of ordered pairs ($\mathbf{v}$, T), such that for some time in (0, T]

$$\mathbf{B}'e^{\mathbf{A}'\tau}\mathbf{v} = \mathbf{0}. \tag{5.18}$$

Lemma 5.1

Given an $n \times m$ matrix $\mathbf{B}$ of rank l and a time τ in R_1^+, there exists a set of $\mathbf{v}$'s in ∂S of dimension $n - l - 1$, such that Equation 5.18 holds.

Proof:

Since $e^{\mathbf{A}'\tau}$ has rank n for all $\mathbf{A}'$ and finite τ and since B has rank l, the matrix $\mathbf{B}'e^{\mathbf{A}'\tau}$ has rank $l \leq n$. Then, the null space[1] of $\mathbf{B}'e^{\mathbf{A}'\tau}$ has dimension $n - l$. Since we are only concerned with these $\mathbf{v}$'s in ∂S, the set has dimension $n - l - 1$.

As τ ranges over R_1, the set of $\mathbf{v}$'s such that the element $(\mathbf{v}, \tau)$ satisfies Equation 5.18 traces out a subset S_q on ∂S of dimension $n - l$. The set of all trajectories which terminate on S_q will form an $(n - l + 1)$ dimensional hypersurface in C/S. We conclude, then, that

[1] See, for example, Reference 41.

Q has dimension n — l + 1. Since l is at least 2, Q has dimension of at most n — 1 (as was the case in Problem Z_1). However, if **B** has rank n, Q must be empty, since S_q will be empty for all τ.

Theorem 5.1

The function **M** has continuous partial derivatives of all orders everywhere on D/Q.

Proof:

The first and second partial derivatives of **M** are examined here, and it is deduced that the higher derivatives also satisfy the theorem. From Equation 5.14, we have

$$\frac{\partial \mathbf{M}}{\partial T} = -\mathbf{AM}(\mathbf{v}, T) + \frac{\mathbf{BF}'(T)\mathbf{v}}{\|\mathbf{F}'(T)\mathbf{v}\|_{\mathbf{R}}}, \tag{5.19}$$

$$\frac{\partial^2 \mathbf{M}}{\partial T^2} = \mathbf{A}^2\mathbf{M}(\mathbf{v}, T) - \frac{\mathbf{ABF}'(T)\mathbf{v}}{\|\mathbf{F}'(T)\mathbf{v}\|_{\mathbf{R}}} + \frac{\mathbf{BB}'\mathbf{A}e^{\mathbf{A}'T}\mathbf{v}}{\|\mathbf{F}'(T)\mathbf{v}\|_{\mathbf{R}}}$$

$$- \frac{\langle \mathbf{F}'(T)\mathbf{v}, \mathbf{BA}'e^{\mathbf{A}'T}\mathbf{v}\rangle \mathbf{BF}'(T)\mathbf{v}}{\|\mathbf{F}'(T)\mathbf{v}\|_{\mathbf{R}}^3},$$

$$\frac{\partial \mathbf{M}}{\partial v_i} = e^{-\mathbf{A}T}\left\{\mathbf{i} + \int_0^T \left(\frac{\mathbf{F}(s)\mathbf{F}'(s)\mathbf{i}}{\|\mathbf{F}'(s)\mathbf{v}\|_{\mathbf{R}}} - \frac{\mathbf{F}(s)\mathbf{F}'(s)\mathbf{v}}{\|\mathbf{F}'(s)\mathbf{v}\|_{\mathbf{R}}^3}\right.\right.$$

$$\left.\left.\langle \mathbf{F}'(s)\mathbf{i}, \mathbf{F}'(s)\mathbf{v}\rangle\right) ds\right\}; \tag{5.20}$$

$$\frac{\partial^2 \mathbf{M}}{\partial v_j \partial v_i} = e^{-\mathbf{A}T} \int_0^T \left\{ - \frac{\mathbf{F}(s)\mathbf{F}'(s)\mathbf{i}}{\|\mathbf{F}'(s)\mathbf{v}\|_{\mathbf{R}}^3} \langle \mathbf{F}'(s)\mathbf{j}, \mathbf{F}'(s)\mathbf{v}\rangle \right.$$

$$- \frac{\langle \mathbf{F}'(s)\mathbf{i}, \mathbf{F}'(s)\mathbf{v}\rangle}{\|\mathbf{F}'(s)\mathbf{v}\|_{\mathbf{R}}^3} \mathbf{F}(s)\mathbf{F}'(s)\mathbf{j}$$

$$- \frac{\langle \mathbf{F}'(s)\mathbf{i}, \mathbf{F}'(s)\mathbf{j}\rangle}{\|\mathbf{F}'(s)\mathbf{v}\|_{\mathbf{R}}^3} \mathbf{F}(s)\mathbf{F}'(s)\mathbf{v}$$

$$\left. + \frac{3\langle \mathbf{F}'(s)\mathbf{i}, \mathbf{F}'(s)\mathbf{v}\rangle}{\|\mathbf{F}'(s)\mathbf{v}\|_{\mathbf{R}}^5} \mathbf{F}(s)\mathbf{F}'(s)\mathbf{v} \right\} ds, \tag{5.21}$$

$$\frac{\partial^2 \mathbf{M}}{\partial T \partial v_i} = -\mathbf{A}\frac{\partial \mathbf{M}}{\partial v_i} + e^{-\mathbf{A}T}\frac{\mathbf{F}(T)\mathbf{F}'(T)}{\|\mathbf{F}'(t)\mathbf{v}\|_{\mathbf{R}}}$$

$$\left(\mathbf{i} - \frac{\langle \mathbf{F}'(T)\mathbf{i}, \mathbf{F}'(T)\mathbf{v}\rangle}{\|\mathbf{F}'(T)\mathbf{v}\|_{\mathbf{R}}^2}\right), \tag{5.22}$$

where **i** is an n-vector with all zero elements except the ith which is 1, and the same holds true for **j**. Since, if $(\mathbf{v}, T) \notin Q$, $\|\mathbf{F}'(s)\mathbf{v}\|_{\mathbf{R}} \neq 0$ for s in $[0, T]$, it follows that $\partial \mathbf{M}/\partial T$, $\partial^2 \mathbf{M}/\partial T^2$, $\partial \mathbf{M}/\partial v_i$, $i = 1, 2, \ldots, n$, $\partial^2 \mathbf{M}/\partial v_i \partial v_j$, $i = 1, 2, \ldots, n$, $j = 1, 2, \ldots, n$, and $\partial^2 \mathbf{M}/\partial T \partial v_i$, $i = 1, 2, \ldots, n$ exist and are continuous on D/Q. The higher-order derivatives follow similarly.

Definition 5.3

Let $\mathbf{M}_{\mathbf{v}}(\mathbf{v}, T)$ denote the partial derivative of $\mathbf{M}(\mathbf{v}, T)$ with respect to **v** in the sense of Frechet, that is,

$$\mathbf{M}_{\mathbf{v}}(\mathbf{v}, T) = e^{-\mathbf{A}T}[I + E], \tag{5.23}$$

where

$$\mathbf{E} \equiv \int_0^T \frac{\mathbf{F}(s)\mathbf{F}'(s)}{\|\mathbf{F}'(s)\mathbf{v}\|_{\mathbf{R}}} - \frac{\mathbf{F}(s)\mathbf{F}'(s)\mathbf{v}\rangle\langle\mathbf{F}(s)\mathbf{F}'(s)\mathbf{v}}{\|\mathbf{F}'(s)\mathbf{v}\|_{\mathbf{R}}^3}\, ds. \tag{5.24}$$

Definition 5.4 $[\mathbf{G}(\mathbf{v}, T)]$

Let $(\mathbf{v}, T) \in D/Q$ and let $\theta = (1/r)\mathbf{h}^{-1}(\mathbf{v})$. Then, the first derivative of **M** (in the sense of Frechet), restricted to D, and with respect to all arguments, is an $n \times n$ matrix denoted by $\mathbf{G}(\mathbf{v}, T)$ or, more simply, by **G**, and we have

$$\mathbf{G}(\mathbf{v}, T) = [r\mathbf{M}_{\mathbf{v}}(\mathbf{v}, T)\mathbf{h}_\theta(\theta) \mid \mathbf{M}_T(\mathbf{v}, T)], \tag{5.24}$$

$$\theta = \frac{1}{r}\mathbf{h}^{-1}(\mathbf{v}). \tag{5.26}$$

Here $\mathbf{M}_T(\mathbf{v}, T)$ is the first partial derivative of **M** with respect to T at $(\mathbf{v}, T)$.

As in Problem Z_1, the matrix $\mathbf{G}(\mathbf{v}, T)$ exists almost everywhere on D. However, there are two differences here. The first is that Q is composed entirely of elements of the Q_μ type in that $\mathbf{G}(\mathbf{v}, T)$ "blows up" on Q. The second difference is that Q can be empty (if **B** has rank n) in Problem Z_2. Also, the number of surfaces comprising Q is smaller in Problem Z_2 when $l \geq 2$; hence, the set Q should give less trouble. In other words, $\mathbf{M}(\mathbf{v}, T)$ is "smoother" in Problem Z_2. However, Problem Z_2 has computational disadvantages, because the evaluation of $\mathbf{M}(\mathbf{v}, T)$ is not a simple numerical problem by any means. Neverthe less, approximate methods are available for evaluating integrals. Using such methods, one could probably ignore the set Q altogether.

Theorem 5.2

Let $(\mathbf{v}, T) \in D/Q$. Then, the $n \times n$ matrix $\mathbf{G}(\mathbf{v}, T)$ has rank n if and only if

$$-\langle \mathbf{v}, \mathbf{A}\mathbf{v}\rangle + \|\mathbf{B}\mathbf{v}\|_{\mathbf{R}} > 0. \tag{5.27}$$

Proof:

Following the proof of Theorem 4.5, one must show that the matrix **E** of Equation 5.24 is positive semidefinite. The remainder of the proof follows precisely that of Theorem 4.5. Obviously, **E** is positive semidefinite if

$$\frac{\mathbf{x}'\mathbf{F}(s)\mathbf{F}'(s)\mathbf{x}}{\|\mathbf{F}'(s)\mathbf{v}\|_{\mathbf{R}}} \geqslant \frac{\mathbf{x}'\mathbf{F}(s)\mathbf{F}'(s)\mathbf{v}\rangle\langle\mathbf{x}'\mathbf{F}(s)\mathbf{F}'(s)\mathrm{v}}{\|\mathbf{F}'(s)\mathbf{v}\|_{\mathbf{R}}^{3}} \tag{5.28}$$

for all s in $[0, T]$.

It can be seen immediately that **E** is not definite, since Relation 5.28 holds with the equality for all s if $\mathbf{x} = \mathbf{v}$. Rewriting Equation 5.28 and canceling out $\|\mathbf{F}'(s)\mathbf{v}\|_{\mathrm{R}} \neq 0$, we have

$$\|\mathbf{F}'(s)\mathbf{x}\|_{\mathbf{R}}^{2}\|\mathbf{F}'(s)\mathbf{v}\|_{\mathbf{R}}^{2} \geqslant \langle\mathbf{F}'(s)\mathbf{x}, \mathbf{R}\mathbf{F}'(s)\mathbf{v}\rangle^{2}. \tag{5.29}$$

But Relation 5.29 is precisely Schwartz's Inequality for real n-vectors; that is, we let

$$\mathbf{z} = \mathbf{F}'(s)\mathbf{x}, \tag{5.30}$$

$$\mathbf{y} = \mathbf{F}'(s)\mathbf{v}. \tag{5.31}$$

Then Equation 5.29 implies that

$$\|\mathbf{z}\|_{\mathrm{R}}\|\mathbf{y}\|_{\mathrm{R}} \geqslant |\langle\mathbf{z}, \mathbf{R}\mathbf{y}\rangle|, \tag{5.32}$$

which is true (Schwartz) for all $\mathbf{z}, \mathbf{y} \in R_n$. Then, $\mathbf{G}^{-1}(\mathbf{v}, T)$ exists everywhere on D/Q if and only if Relation 5.27 holds (see Theorem 4.5).

The modification of the problem that replaces the origin as a target with the hypersphere again guarantees that **G** is nonsingular whenever it exists, provided that Relation 5.27 holds. It may not be necessary to modify the problem to ensure that the matrix **G** has rank n (that is, it is not necessary that $\mathbf{M_v}$ be definite). However, in Problem Z_1, the fact that **G** was singular in the unmodified problem was closely related to the fact that the set $\partial\psi(T, \mathbf{x}_0)$ had "corners" (see Figure 4.2). These corners exist because the time-optimal control is discontinuous (switchings). If $\mathbf{u} \in \Omega_b$, the time-optimal controls (some of them) are also discontinuous (Lemma 5.1), unless $\mathbf{B}$ has rank n. Hence, it is conjectured that in the unmodified problem, $\mathbf{G}(\mathbf{v}, T)$ will be singular for some $(\mathbf{v}, T)$ if $\mathbf{B}$ does not have rank n.

5.4 The Iterative Procedure

The iterative procedure for Problem Z_2 is described in Section 4.4, and the comments given in that section apply equally well to Problem Z_2.

5.5 Convergence

Section 4.5 describes the convergence properties of the iterative procedure. However, a few points can be made in comparing Problems Z_2 and Z_1.

If $\mathbf{B}$ has rank n and Relation 5.27 holds, then $\mathbf{G}(\mathbf{v}, T)$, $\mathbf{G}^{-1}(\mathbf{v}, T)$, and $\mathbf{M}^2(\mathbf{v}, T)$ will exist everywhere. Hence, one can apply Theorem 4.7 and conclude that the iterative procedure will converge monotonically to the solution $(\mathbf{v}^*, T^*)$ for any initial guess, such that $T_k \in R_1^+$ for all k. This condition is easy to satisfy by using the property

$$\mathbf{M}(\mathbf{v}, T) = -\mathbf{M}(-\mathbf{v}, T). \tag{5.33}$$

That is, one can check after each iteration to see whether

$$\|\mathbf{e}(-\mathbf{v}_k, T_k)\| < \|\mathbf{e}(\mathbf{v}_k, T_k)\|, \tag{5.34}$$

and if it is, one can change the sign of $\mathbf{v}_k$. An inspection of Figure 4.6 will show that if $\dot{\mathbf{x}}(\mathbf{v}_k, T_k)$ is such that the sequence $\{T_k\}$ must go negative to ensure that

$$\|\mathbf{e}_{k+1}\| < \|\mathbf{e}_k\|, \tag{5.35}$$

then the current guess $\mathbf{v}_k$ is on the "wrong side" of the hypersphere and can be corrected by choosing $(-\mathbf{v}_k, T_k)$ as the proper guess.

If $\mathbf{B}$ does not have rank n, then the iteration may encounter the set Q. However, the low dimensionality of Q will ensure that this set causes little difficulty.

5.6 The Initial-Guess Procedure

The initial-guess procedure of Section 4.6 also applies to Problem Z_2. The equivalent equation to Equation 4.112 becomes

$$\langle \mathbf{v}^*, e^{\mathbf{A}T^*}\mathbf{x}_0 \rangle = r^2 + \int_0^T \|\mathbf{F}'(s)\mathbf{v}\|_{\mathbf{R}}\, ds. \tag{5.36}$$

The problem can be separated in exactly the same manner as that in

Section 4.6, in which case the vectors $\mathbf{b}_i$ in Equations 4.117 become matrices $\mathbf{B}_i$.

It was assumed in this chapter that the matrix $\mathbf{Q}$ of Chapter 4 was the $n \times n$ identity matrix. This assumption was not necessary. Similarly, it should be obvious from Section 4.8 that the time-varying case can also be treated in this manner.

No numerical results were obtained for this class of problems.

6. The Minimum-Fuel Control Problem

6.1 Introduction

The computation of fixed-time fuel-optimal controls (that is, Problem Z_3 of Definition 1.13) is studied in this chapter. There are many similarities between this problem and problem Z_1. Consequently, wherever possible, the reader is referred to an appropriate proof in Chapter 4. Problem Z_3 is briefly restated in Section 6.2 where function $\mathbf{M}(\mathbf{v}, \alpha)$ is also defined. Section 6.3 contains preliminary theorems and definitions establishing the properties of $\mathbf{M}$ (that is, the sets Q and D, the differentiability of $\mathbf{M}$, etc.) that are pertinent to the iterative procedure. This procedure is introduced formally in Section 6.4. In Section 6.5, it is stated that the convergance arguments of Section 4.5 apply to the iterative procedure of Section 6.4 if T_k is replaced by α_k. This point is illustrated by proving the equivalent of Lemma 4.1 for Problem Z_3. The problem of making an initial guess is treated in Section 6.6. Finally, the multi-input and time-varying cases are discussed in Section 6.7.

6.2 The Function M

Problem Z_3 is briefly restated (see Definition 1.13) for orientation. Given the following:

$$\dot{\mathbf{x}}(t) = \mathbf{A}\mathbf{x}(t) + \mathbf{b}u(t), \qquad \mathbf{x}(0) = \mathbf{x}_0, \tag{6.1}$$

$$S = \{\mathbf{x}: \langle \mathbf{x}, \mathbf{x} \rangle - r^2 \leq 0\}, \qquad r > 0, \tag{6.2}$$

$$J = \int_0^T |u| \, dt, \qquad |u| \leq 1, \tag{6.3}$$

$$T \text{ is fixed and } T > T^*, \tag{6.4}$$

$$\mathbf{x}_0 \in C/S \text{ (see Definition 1.6).} \tag{6.5}$$

Find the control that steers the state from the initial condition $\mathbf{x}_0$ to an unspecified terminal state $\mathbf{v} \in S$, such that the elapsed time is T and the cost functional J is minimized. As before, the Minimum Principle is applied and one concludes that if the optimal terminal state $\mathbf{v}^*$ is in the boundary of S (that is, $\mathbf{v}^* \in \partial S$), then the boundary condition on the costate differential system is (see NC10, Theorem 1.2)

$$\mathbf{p}^*(T) = \alpha^* \mathbf{v}^*, \qquad \alpha^* > 0. \tag{6.6}$$

Also, if the optimal terminal state lies in the interior of S (that is, $\mathbf{v}^* \in S/\partial S$), then (see NC11, Theorem 1.2), we have

$$p^*(T) = \mathbf{0}. \tag{6.7}$$

The Hamiltonian for Problem Z_3 is

$$H = p_0 \mid u(t) \mid + \langle \mathbf{p}(t), \mathbf{A}\mathbf{x}(t) \rangle + \langle \mathbf{b}, \mathbf{p}(t) \rangle u(t), \tag{6.8}$$

from which we obtain

$$\dot{\mathbf{p}}^*(t) = -\mathbf{A}' \, p^*(t), \qquad \mathbf{p}^*(T) = \alpha^* v^*, \qquad \text{or} \qquad \mathbf{p}^*(T) = \mathbf{0}, \tag{6.9}$$

$$\dot{\mathbf{x}}^*(t) = \mathbf{A}\mathbf{x}^*(t) + \mathbf{b}\mathbf{u}^*(t), \qquad \mathbf{x}^*(0) = \mathbf{x}_0, \tag{6.10}$$

and

$$u^*(t) = \begin{cases} -\operatorname{dez}\{\langle \mathbf{p}^*(t), \mathbf{b} \rangle\}, & p_0^* \neq 0, \quad (6.11a) \\ \\ -\operatorname{sgn}\{\langle \mathbf{p}^*(t), \mathbf{b} \rangle\}, & p_0^* = 0, \quad (6.11b) \end{cases}$$

where

$$\operatorname{dez}\{x\} \equiv \begin{cases} +1, & x > 1, \\ -1, & x < -1, \\ 0, & |x| < 1, \\ \text{somewhere in } [0, +1], & x = 1, \\ \text{somewhere in } [0, -1], & x = -1. \end{cases} \tag{6.12}$$

From Equation 6.9, if $\mathbf{p}^*(T) \neq 0$, we have

$$u^*(t) = \begin{cases} -\operatorname{dez}\{\langle \alpha^* \mathbf{v}^*, e^{-\mathbf{A}t}\mathbf{b} \rangle\}, & p_0 \neq 0, \\ \\ -\operatorname{sgn}\{\langle \alpha^* \mathbf{v}^*, e^{-\mathbf{A}t}\mathbf{b} \rangle\}., & p_0 = 0. \end{cases} \tag{6.13}$$

Clearly, for any $\mathbf{v} \neq \mathbf{0}$,

$$\lim_{\alpha^* \to \infty} \text{dez}\{\langle \alpha^* \mathbf{v}^*, e^{-\mathbf{A}t}\mathbf{b}\rangle\} = \text{sgn}\{\langle \alpha^* \mathbf{v}^*, e^{-\mathbf{A}t}\mathbf{b}\rangle\}. \tag{6.14}$$

In fact, if $\langle \mathbf{v}^*, e^{-\mathbf{A}t}\mathbf{b}\rangle$ has no zeros on the interval (0, T), Equation 6.14 holds for α^* greater than some finite determinable number, Hence, if one is prepared to allow α^* to be arbitrarily large, one can include Equation 6.11b in 6.11a and conclude that $p_0^* \neq 0$. Also, from this point of view, one could include Problem Z_1 in this problem by changing relation 6.4 to $T \geq T^*$. This point of view is not adopted, however, because the variable α is used in the proposed iterative procedure and hence must remain finite. On the other hand, since we would like to define the function **M** in terms of a single-control law and an initial-value problem, it is of interest to know if and when $p_0^* \neq 0$.

Lemma 6.1

If the matrix $-\mathbf{A}$ is positive definite and $T > T^*$, where T^* is the minimum time in which the state can be steered from $\mathbf{x}_0 \in C/S$ to ∂S using $|u| \leq 1$, then $p_0^* \neq 0$ in Equation 6.11.

Proof:

Suppose that $p_0^* = 0$. Then, from Equations 6.3 and 6.11b, we see that $J = T$. It is claimed that there is a feasible control for which the cost is less than T. Consider applying the time-optimal control over the interval $(0, T^*]$ and then turning the control off (that is, $u_{(T^*, T]} = 0$). The fuel consumed during this maneuver is less than T, and, since $-\mathbf{A}$ is positive definite, the state will remain in S for $u_{(T^*, T]} = 0$.[1]

Q.E.D.

It is henceforth assumed that the matrix $-\mathbf{A}$ is positive definite and that the controllability matrix **C** is nonsingular. Besides being useful in avoiding the case $p_0^* = 0$, this assumption also ensures, among other things, that Problem Z_3 is normal (see Appendix C and Reference 2).

Lemma 6.1 means that Equation 6.11a describes the control law. It is convenient to state two properties of the dead-zone function.

p1. Unlike sgn $\{ax\}$, the function dez $\{\alpha x\}$ is dependent on α (that is for $\alpha > 0$).

p2. If $\alpha_1 > \alpha_2 > 0$, and if $|\alpha_1 x|, |\alpha_2 x| \neq 1$, then

$$|\text{dez}\{\alpha_1 x\}| \geq |\text{dez}\{\alpha_2 x\}|. \tag{6.15}$$

---

[1] Let $\dot{\mathbf{x}} = \mathbf{A}\mathbf{x}$ (that is, $u = 0$), and let $\mathbf{x}(0) = \mathbf{x}_0$. Then, if $-\mathbf{A}$ is positive definitite, $\|\mathbf{x}(t)\|$ is strictly decreasing, since $(d/dt)\|\mathbf{x}(t)\| = (1/2\|\mathbf{x}(t)\|)\langle \mathbf{x}(t), \mathbf{A}\mathbf{x}(t)\rangle < 0$.

Property p1 means that α must be included in the list of arguments for **M**.

The function $\mathbf{M}(\mathbf{v}, \alpha)$ is now defined operationally in terms of an initial-value problem. Choose $\mathbf{v} \in R_n$ and $\alpha \in R_1^+$, let $\tau = T - t$, and solve the initial-value problem

$$\dot{\mathbf{x}}(\tau) = -\mathbf{A}\mathbf{x}(\tau) + \mathbf{b}\ \mathrm{dez}\{\langle \alpha\, \mathrm{v}, e^{-\mathbf{A}\tau}\mathbf{b}\rangle\}, \qquad \mathbf{x}(\tau = 0) = \mathbf{v} \tag{6.16}$$

to find its solution at $\tau = T$.

Definition 6.1 (M)

Let **M** be the function

$$\mathbf{M}(\mathrm{v}, \alpha) = e^{-\mathbf{A}T}\left\{\mathrm{v} + \int_0^T e^{\mathbf{A}s}\mathbf{b}\ \mathrm{dez}\{\langle \alpha\, \mathrm{v}, e^{\mathbf{A}s}\mathbf{b}\rangle\}\ ds\right\}. \tag{6.17}$$

Note that $\mathbf{M}(\mathrm{v}, \alpha)$ is the solution of Equation 6.16 at $\tau = T$.

6.3 Definitions and Preliminary Theorems

Since $S \times R_1^+$ (that is, $\mathbf{v} \in S$ and $\alpha \in R_1^+$) has dimension $n + 1$, it is of interest to find the conditions under which the optimal terminal point $\mathbf{v}^*$ is in ∂S. One can hardly expect **M** to be one-to-one on a set of dimension greater than its range.

Lemma 6.2

If the terminal state $\mathbf{v}^*$ of a fixed-time fuel-optimal trajectory from some state $\mathbf{x}_0 \in C/S$ is in the interior of S (that is, $\mathbf{v}^* \in S/\partial S$), then the corresponding fixed-time fuel-optimal control is $u^*_{(0,T]} = 0$. (That is the minimum cost $J^* = 0$.)

Proof:

Since the conditions of the Minimum Principle are necessary, $\mathbf{p}^*(T) = 0$ if $\mathbf{v}^* \in S/\partial S$ (Equation 6.7 and/or NC11). Then, from Equation 6.12, we have $\mathbf{p}^*(\tau) = 0$ for all τ, from which, applying NC2 (that is, $p_0^* \neq 0$ if $\mathbf{p}^*(t) = 0$) and Equation 6.11, we obtain $u^*_{(0,T]} = 0$.

Q.E.D.

Lemma 6.1 means that either $\mathbf{v}^* \in \partial S$ or $u^*_{(0,T]} = 0$, or both. This fact is convenient from a computational point of view, for it means that it is necessary to search only the n-dimensional set $\partial S \times R_1^+$. In other words, given any $\mathbf{x}_0 \in C/S$, one can first calculate $e^{\mathbf{A}T}\mathbf{x}_0$, the undriven (that is, homogeneous) response of the plant from $\mathbf{x}_0$ at time T. If $\|e^{\mathbf{A}T}\mathbf{x}_0\| \leq r$, then the solution is the trivial one, name, $u^*_{(0,T]} = 0$. Clearly, from Lemmas 6.1 and 6.2, if $\|e^{\mathbf{A}T}\mathbf{x}_0\| > r$, then $\mathbf{v}^* \in \partial S$. It is henceforth assumed that $\|e^{\mathbf{A}T}\mathbf{x}_0\| > r$.

Lemma 6.3

If $\|e^{\mathbf{A}T}\mathbf{x}_0\| > r$ (that is, if $u^*_{(0,T]} \neq 0$), if the matrix $-\mathbf{A}$ is positive definite, and if $T > T^*$, where T^* is the minimum time in which the state can be steered from $\mathbf{x}_0$ to ∂S, then the fixed-time fuel-optimal control makes at least one switching in the interval $(0, T)$.

Proof:

The proof follows that of Lemma 6.1. Suppose the contrary. Since $u^*_{(0,T]} \neq 0$ (Lemma 6.2), it follws that $J = T$. But $T > T^*$ (see the proof of Lemma 6.1). Q.E.D.

Definition 6.2(D)

Let $D \subseteq \partial S \times R_1^+$ be the set of ordered pairs $(\mathbf{v}, \alpha)$, such that if $(\mathbf{v}, \alpha) \in D$, then $\mathbf{v} \in \partial S$ is the terminal state at $t = T$ of a trajectory from some state in C/S for which $\|e^{AT}\mathbf{x}_0\| > r$ and the conditions of Theorem 1.2 are satisfied

Lemma 6.4

For any given $\mathbf{v} \in \partial S$ there exists a positive real number $\delta(\mathbf{v})$, such that if $\alpha \in [0, \delta(\mathbf{v})]$, the ordered pair $(\mathbf{v}, \alpha) \notin D$.

If $\mathbf{v}$ is such that the function $\langle \mathbf{v}, e^{\mathbf{A}\tau}\mathbf{b}\rangle$ has no zeros in the interval $(0, T)$, then there exists a number $\rho(\mathbf{v})$ such that if $\alpha \geq \rho(\mathbf{v})$, the ordered pair $(\mathbf{v}, \alpha) \notin D$.

Proof:

From Lemma 6.3 and Definition 6.2, we know that a control generated by an ordered pair $(\mathbf{v}, \alpha)$ in D must make at least one switching. We choose

$$\delta(\mathbf{v}) = \frac{1}{\sup\limits_{\tau \in (0,T)} \{|\mathbf{v}, e^{\mathbf{A}\tau}\mathbf{b}\rangle|\}}, \tag{6.18}$$

and

$$\rho(\mathbf{v}) = \frac{1}{\inf\limits_{\tau \in (0,T)} \{|\mathbf{v}, e^{\mathbf{A}\tau}\mathbf{b}\rangle|\}} . \tag{6.19}$$

Q.E.D.

The behavior of a typical $\mathrm{dez}\{\langle \alpha\mathbf{v}, e^{\mathbf{A}\tau}\mathbf{b}\rangle\}$ is illustrated in Figure 6.1 for v fixed.

Lemma 6.4 means that $D \neq \partial S \times R_1^+$. In other words, given a $\mathbf{v} \in \partial S$, the ordered pair $(\mathbf{v}, \alpha)$ will be a candidate for an element of D only if $\delta(\mathbf{v}) < \alpha < \rho(\mathbf{v})$ [from Equations 6.18 and 6.19, $\delta(\mathbf{v}) < \rho(\mathbf{v})$]. Though this fact is a nuisance when programming the computer solution for

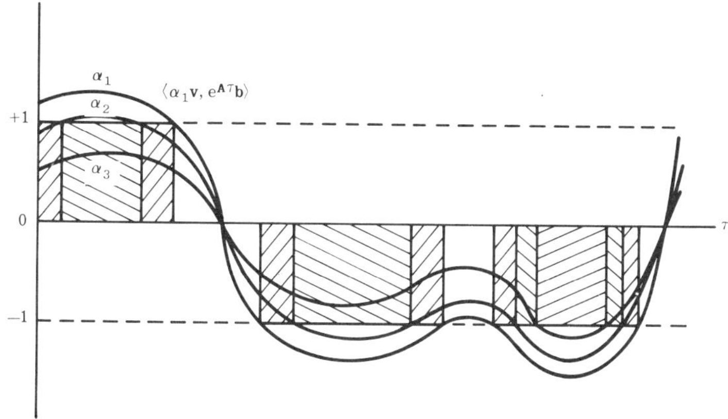

Figure 6.1 Effects of α on dez $\{\langle \alpha\mathbf{v}, e^{\mathbf{A}\tau}\mathbf{b}\rangle\}$ $(\alpha_1 > \alpha_2 > \alpha_3)$.

Problem Z_3, it does not make the approach impractical, because one does not have to calculate $\rho(\mathbf{v})$ and $\delta(\mathbf{v})$ in advance. One can simply increase α if the current guess yields a control with $u_{(0,T]} = 0$ or decrease α if $u_{(0,T]} = \pm 1$ until a control with a switching is generated.

Theorem 6.1

If the matrix $-\mathbf{A}$ is positive definite, then $D = \partial S \times (\delta(\mathbf{v}), \rho(\mathbf{v}))$.

Proof:

The conditions of the Minimum Principle are both necessary and sufficient for Problem Z_3 ([25] Theorem C). By construction, any trajectory generated by any ordered pair in D satisfies the applicable conditions of the Minimum Principle (see Theorem 1.2). Then, from Definition 6.2, one need show only that both $\mathbf{M}(v, \alpha)$ and $e^{\mathbf{A}T}\mathbf{M}(\mathbf{v}, \alpha) \in C/S$. Since $-\mathbf{A}$ is positive definite, $\|e^{\mathbf{A}T}\mathbf{x}\| < \|\mathbf{x}\|$ for all $\mathbf{x}$ in R_n (see the proof of Lemma 6.1). It follows that $\mathbf{M}(\mathbf{v}, \alpha) \in C/S$ if $e^{\mathbf{A}T}\mathbf{M}(\mathbf{v}, \alpha) \in C/S$. Now, from Equation 6.17, we have

$$\|e^{\mathbf{A}T}\mathbf{M}(\mathbf{v}, \alpha)\| = \|\mathbf{v} + \int_0^T \langle \alpha\mathbf{v}, e^{\mathbf{A}s}\mathbf{b}\,\mathrm{dez}\{\langle \alpha\mathbf{v}, e^{\mathbf{A}s}\mathbf{b}\rangle\}\, ds\,\|$$

$$= [\,\|\mathbf{v}\|^2 + \frac{2}{\alpha}\int_0^T \langle \alpha\mathbf{v}, e^{\mathbf{A}s}\mathbf{b}\rangle\, \mathrm{dez}\{\langle \alpha\mathbf{v}, e^{\mathbf{A}s}\mathbf{b}\rangle\}\, ds$$

$$+ \|\int_0^T e^{\mathbf{A}s}\mathbf{b}\; \mathrm{dez}\{\langle \alpha\mathbf{v}, e^{\mathbf{A}s}\mathbf{b}\rangle\}\|^2]^{1/2}. \tag{6.20}$$

But $\|\mathbf{v}\|^2 = r^2$, and if $\alpha > \delta(\mathbf{v})$, each of the integral terms under the radical in Equation 6.20 is positive. It follows that $\|e^{\mathbf{A}T}\mathbf{M}(\mathbf{v}, \alpha)\| > r$.

Q.E.D.

Theorem 6.2

The function $\mathbf{M}(\mathbf{v}, \alpha)$ is a one-to-one continuous function on D, and the function $J(\mathbf{v}, \alpha)$ is continuous and strictly increasing in α for fixed $\mathbf{v}$.

Proof:

The continuity of $J(\mathbf{v}, \alpha)$ follows if $\mathbf{M}(\mathbf{v}, \alpha)$ is continuous. The continuity of $\mathbf{M}(\mathbf{v}, \alpha)$ is proved in Section V of Reference 34. Athans [2] has shown that the extremal controls are unique. It follows that $\mathbf{M}(\mathbf{v}, \alpha)$ is 1:1 on D.

Now, we have

$$J(\mathbf{v}, \alpha) = \int_0^T |\operatorname{dez}(\langle \alpha \mathbf{v}, e^{\mathbf{A}s}\mathbf{b}\rangle)| \, ds, \tag{6.21}$$

and from Equation 6.15, we obtain

$$|\operatorname{dez}(\langle \alpha_1 \mathbf{v}_1, e^{\mathbf{A}s}\mathbf{b}\rangle)| \geq |\operatorname{dez}(\langle \alpha_2 \mathbf{v}_2, e^{\mathbf{A}s}\mathbf{b}\rangle)| \tag{6.22}$$

everywhere in $(0, T)$ if $\alpha_1 > \alpha_1$ and $\alpha_2 \in (\delta)\mathbf{v}), \rho(\mathbf{v}))$, there is at least one switching and, hence, at least one nonzero interval over which the strict inequality holds in Equation 6.22.

Q.E.D.

In other words, given a $\mathbf{v}$, the function $J(\mathbf{v}, \alpha)$, considered as a function of α, is 1:1 on $(\delta(\mathbf{v}), \rho(\mathbf{v}))$.

As in the time-optimal problems, the differentiability of $\mathbf{M}(\mathbf{v}, \alpha)$ and, in this case, $J(\mathbf{v}, \alpha)$ is investigated. One proceeds in a similar manner, and a representation for $\mathbf{M}$ is found that displays the set Q on which the derivatives do not exist. Two "typical" fuel-optimal controls are illustrated in Figure 6.2. As illustrated, a fuel-optimal control alternates between the levels +1, 0, and −1. Such a control can be completely characterized by four parameters: (1) the initial value of the control, (2) the number of switchings, (3) the ordered set of switch times, and (4) the values of $\langle \alpha\mathbf{v}, e^{\mathbf{A}\tau}\mathbf{b}\rangle$ at the switch times.

Definition 6.3 ($\eta(\mathbf{v}, \alpha)$)

Let $\eta(\mathbf{v}, \alpha)$ be the piecewise constant function from $R_n \times R_1$ into R_1:

$$\eta(\mathbf{v},\alpha) = \begin{cases} 1, & |\langle \alpha\mathbf{v}, \mathbf{b}\rangle| < 1 \\ 1, & |\langle \alpha\mathbf{v}, \mathbf{b}\rangle| = 1, \quad |\langle \alpha\mathbf{v}, e^{\mathbf{A}(\tau=0^+)}\mathbf{b}\rangle| < 1, \\ 2, & |\langle \alpha\mathbf{v}, \mathbf{b}\rangle| > 1 \\ 2, & |\langle \alpha\mathbf{v}, \mathbf{b}\rangle| = 1, \quad |\langle \alpha\mathbf{v}, e^{\mathbf{A}(\tau=0^+)}\mathbf{b}\rangle| > 1. \end{cases} \tag{6.23}$$

This function is defined for "book-keeping" purposes. Two "typical" controls are illustrated in Figure 6.2. One arbitrarily chooses $\eta(\mathbf{v}, \alpha) = 1$ if $u(0^+) = 0$ and $\eta(\mathbf{v}, \alpha) = 2$ if $u(0^+) = \pm 1$.

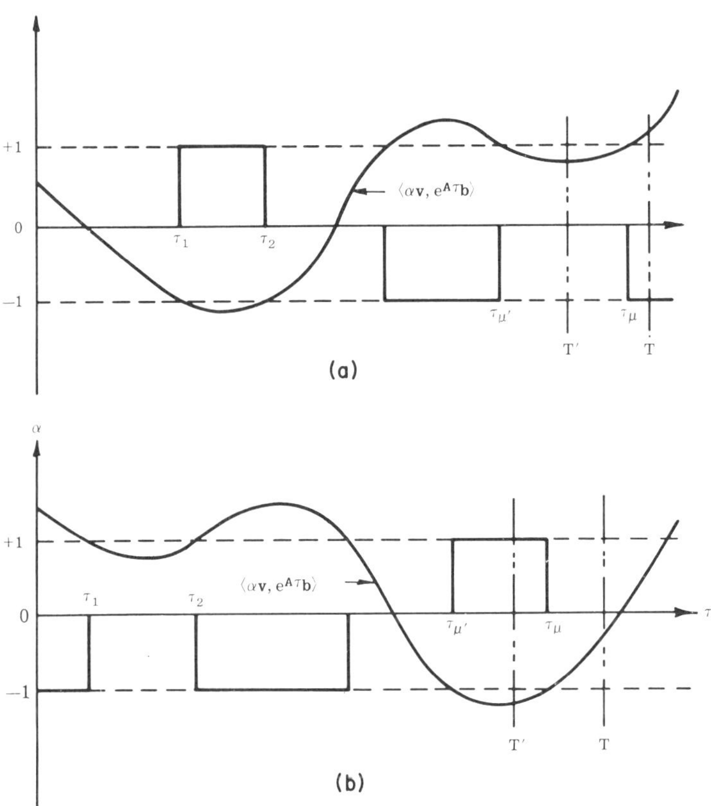

Figure 6.2 Illustration of type-1 and type-2 control.

Definition 6.4 (Q)

Let $Q \subset D$ be the set of ordered pairs $(\mathbf{v}, \alpha)$, such that

$$|\langle \alpha\mathbf{v}, \mathbf{b}\rangle| = 1. \qquad (6.24)$$

Equation 6.24 defined two hyperplanes in R_n, neither of which contain the origin. Clearly, given any $\alpha > 0$, there exists a cone of $\mathbf{v}$'s in ∂S, symmetric about $\mathbf{b}$, of dimension $n - 2$, such that $(\pm\mathbf{v}, \alpha) \in \mathbf{Q}_\eta$. Hence, Q has dimension $n - 1$.

Definition 6.5 ($\mu(\mathbf{v}, \alpha)$)

Let $\mu(\mathbf{v}, \alpha)$ be an integer-valued function on D such that the control $u_{(0,T)}$ generated by the ordered pair $(\mathbf{v}, \alpha)$ makes exactly $\mu(\mathbf{v}, \alpha)$ switchings in the open interval (0, T).

From Lemma 6.3, $\mu(\mathbf{v}, \alpha) \geq 1$ on D.

Definition 6.6 ($\tau_j(\mathbf{v}, \alpha), j = 1, 2, \ldots, \mu(\mathbf{v}, \alpha)$)

The switch times $\tau_j(\mathbf{v}, \alpha), j = 1, 2, \ldots, \mu(\mathbf{v}, \alpha)$ are a set of ordered real numbers such that

1. $\tau_j(\mathbf{v}, \alpha) \in (0, T),$ (6.25)

2. $|\langle \alpha \mathbf{v}, e^{\mathbf{A}\tau_j(\mathbf{v},\alpha)}\mathbf{b}\rangle| = 1,$ (6.26)

3. $\text{sgn}\{|\langle \alpha \mathbf{v}, e^{\mathbf{A}\tau_j^+(\mathbf{v},\alpha)}\mathbf{b}\rangle| - 1\} = -\,\text{sgn}\{|\alpha \mathbf{v}, e^{\mathbf{A}\tau_j^-(\mathbf{v},\alpha)}\mathbf{b}\rangle| - 1\},$ (6.27)

4. $\tau_j(\mathbf{v},\alpha) < \tau_{j+1}(\mathbf{v}, \alpha)$ for all j. (6.28)

See Definition 4.6.

Definition 6.7

Let $Q_\mu \subset D$ be a set of ordered pairs $(\mathbf{v}, \alpha)$ such that for one or more times τ in (0, T] the equations

$$|\langle \alpha \mathbf{v}, e^{\mathbf{A}\tau}\mathbf{b}\rangle| = 1, \tag{6.29}$$

$$|\langle \alpha \mathbf{v}, e^{\mathbf{A}\tau}\mathbf{b}\rangle = 0, \tag{6.30}$$

hold simultaneously.

Suppose that n = 2. Since $e^{\mathbf{A}\tau}$ and $\mathbf{C} = [\mathbf{b}, \mathbf{Ab}]$ are nonsingular, given any τ there exists a unique $\mathbf{v}_q$ and α_q such that $\pm\mathbf{v}_q$, α_q, and τ satisfy Equations 6.29 and 6.30 with $\mathbf{v}_q \in \partial S$. As τ ranges over (0, T), the ordered pairs $(\pm\mathbf{v}_q(\tau), T_q(\tau))$ trace out two curves (sets of dimension one) in the two-dimensional space D. In the general n-dimensional case, Q_μ has dimension n − 1.

Definition 6.8 (Q_T)

Let $Q_T \subset D$ be a set of ordered pairs $(\mathbf{v}, \alpha)$ such that

$$|\langle \alpha \mathbf{v}, e^{\mathbf{A}T}\mathbf{b}\rangle| = 1. \tag{6.31}$$

See Definition 6.4 in order to compare Q_η and Q_T. Here Q_T has dimension n − 1.

Definition 6.9 (Q)

Let Q denote the union

$$Q = Q_\mu \cup Q_\eta \cup Q_T; \tag{6.32}$$

that is, Q is a collection of several (n — 1)-dimensional hypersurfaces in D.

Definition 6.10 (D_i)

The set D/Q is the union (perhaps infinite) of open sets D_i over which the function $\eta(\mathbf{v}, \alpha)$ and $\mu(\mathbf{v}, \alpha)$ are constant, and on which, if

$$|\langle \alpha\,\mathbf{v}, e^{A\hat{\tau}}\mathbf{b}\rangle| = 1, \tag{6.33}$$

then

$$\langle \alpha\mathbf{v}, \mathbf{A}e^{\mathbf{A}\hat{\tau}}\mathbf{b}\rangle \neq 0. \tag{6.34}$$

Example 6.1. Let the plant have the form

$$\dot{x}_1 = \lambda_1(x_1 + u), \tag{6.35}$$

$$\dot{x}_2 = \lambda_2(x_2 + u), \tag{6.36}$$

where λ_1, λ_2 are assumed real and negative, and let $v_1 = r\cos\theta$ and $v_2 = r\sin\theta$, so that D may be represented in the $\theta \sim \alpha$ plane. Then, from Equation 6.24, we have

$$Q_\eta = \{(\theta,\alpha);\ |\alpha r(\lambda_1 \cos\theta + \lambda_2 \sin\theta)| = 1\}, \tag{6.37}$$

and, from Equation 6.31,

$$Q_T = \{(\theta, \alpha);\ |\alpha r(\lambda_1 e^{\lambda_1 T}\cos\theta + \lambda_2 e^{\lambda_2 T}\sin\theta)| = 1\}, \tag{6.38}$$

replacing Equation 6.29 by

$$\langle \alpha\mathbf{v}, e^{A\tau}\mathbf{b}\rangle = \Delta, \qquad \Delta = \pm 1. \tag{6.39}$$

Solving Equation 6.30 and using Equation 6.39, we obtain

$$Q_\mu \quad = \{(\theta, \alpha);\ \alpha\cos\theta = \frac{\Delta\lambda_2 e^{-\lambda_1 t}}{r\lambda_1(\lambda_2 - \lambda_1)},$$

$$\alpha\sin\theta = \frac{\Delta\lambda_1 e^{-A_2 t}}{r\lambda_2(\lambda_1 - \lambda_2)}, \qquad \Delta = \pm 1, t \in (0, T)\}. \tag{6.40}$$

In this example, Q_μ is actually empty because the α satisfying Equation 6.40 is such that $u_{(0,T]} = 0$ (that is, $\alpha = \rho(\mathbf{v})$). These sets are illustrated in Figure 6.3 along with the sets D_1 for $\lambda_1 = -1$, $\lambda_2 = -2$, $r = 1$, and $T = 3$.

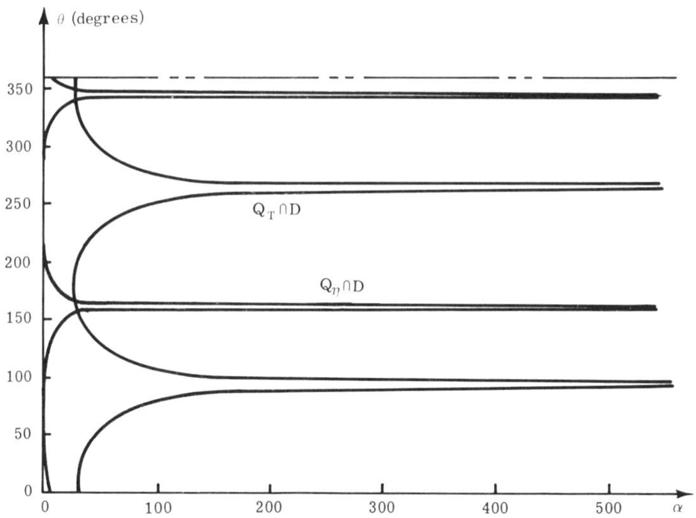

Figure 6.3 Sets Q_T and Q_η for a second-order plant.

Definition 6.11 ($\omega_j(\mathbf{v}, \alpha)$)

Let

$$\omega_j(\mathbf{v},\alpha) = \begin{cases} u(0^+), & j = 0, \\ \langle \alpha\mathbf{v}, e^{\mathbf{A}\tau_j(\mathbf{v},\alpha)}\mathbf{b}\rangle, & j = 1, 2, \ldots, \mu(\mathbf{v}, \alpha), \\ u(T), & j = \mu(\mathbf{v}, \alpha) + 1, \end{cases} \tag{6.41}$$

where the $\tau_j(\mathbf{v}, \alpha)$, $j = 1, 2, \ldots, \mu(\mathbf{v}, \alpha)$ are the switch times.

Properties of $\omega_j(\mathbf{v}, \alpha)$

1. $\omega_j(\mathbf{v}, \alpha) = \pm 1, \quad j = 1, 2, \ldots, \mu(\mathbf{v},\alpha),$ (6.42)

2. $\omega_0(\mathbf{v}, \alpha) = \omega_1(\mathbf{v}, \alpha)\ (\eta - 1),$ (6.43)

3. $\omega_{\mu+1}(\mathbf{v}, \alpha) = \omega_\mu(\mathbf{v},\alpha)\ \dfrac{(1 + (-1)^{\mu+\eta})}{2},$ (6.44)

4. $\omega_{2l+2-\eta}(\mathbf{v},\alpha) = \omega_{2l+3-\eta}(\mathbf{v},\alpha), \quad l = 0, 1, \ldots,$ (6.45)

5. $$\frac{\partial \omega_j(\mathbf{v},\alpha)}{\partial \alpha} = 0 \quad \text{for all j,} \quad (\mathbf{v}, \alpha) \in D_0\{\mathbf{M}\}/\mathbf{Q}, \tag{6.46}$$

6. $$\frac{\partial \omega_j(\mathbf{v},\alpha)}{\partial v_i} = 0 \quad \text{for all i and all j,} \quad (\mathbf{v}, \alpha) \in D_0\{\mathbf{M}\}/\mathbf{Q}, \tag{6.47}$$

where η means $\eta(\mathbf{v}, \alpha)$ and $\mu(\mathbf{v}, \alpha)$.

In words, $\omega_j(\mathbf{v}, \alpha)$ is +1 if the control at the jth switch time is switching either from −1 to 0 or from 0 to −1, and it is −1 if the control is switching from +1 to 0 or vice versa. Properties 2, 3, and 4 can best be seen by examining Figure 6.2. Property 2 means that when $\eta = 1$ (a type-1 control), the control is zero until the first switching, and when $\eta = 2$ (a type-2 control), the control is ± 1 for τ in $(0, \tau_1(\mathbf{v}, \alpha))$. Property 3 accounts for the control in the interval $(\tau_\mu(\mathbf{v}, \alpha), T)$ in the same manner. Obviously, whether $u(\tau = T)$ is ± 1 or 0 depends on the number of switchings μ and the control type η (see Figure 1.4). Property 4 accounts for the fact that the control is constant between the switch times. For example in Figure 6.2a $\eta = 1$, and hence $\omega_0 = 0$ from Property 4, $\omega_1 = \omega_2, \omega_3 = \omega_4$, and using $T = T'$, we find that $\omega_{\mu+1} = 0$ from Property 3. Properties 5 and 6 follow for $(\mathbf{v}, \alpha) \in D$ from Equation 6.42 and from the fact that on each set D_i the number of switchings $\mu(\mathbf{v},\alpha)$ and the control type $\eta(\mathbf{v},\alpha)$ are constants.

Theorem 6.3

The function $\mathbf{M}(v,\alpha)$ and $J(\mathbf{v}, \alpha)$ (see Lemma 6.5) have continuous partial derivatives of all orders on every open set $D_i \subset D$.

Proof:

Since $(\mathbf{v}, \alpha)$ ($\in D_i$ for some i), the two functions $\mathbf{M}(\mathbf{v}, \alpha)$ and $J(\mathbf{v}, \alpha)$ can be written in a form in which the derivatives can be conveniently taken. Thus, we have

$$J(\mathbf{v},\alpha) = \int_0^T |u(\tau)|\, d\tau = \sum_{l=0}^{\xi} \int_{\tau_{2l+2-\eta}}^{\tau_{2l+3-\eta}} d\tau, \tag{6.48}$$

where the τ_j (the switch times) are written without arguments for convenience, and where

$$\tau_0 \equiv 0, \tag{6.49}$$

$$\tau_{\mu+1} \equiv T, \tag{6.50}$$

and

$$2\xi + 3 - \eta = \mu + \frac{(1 + (-1)^{\mu+\eta})}{2}, \tag{6.51}$$

wherein μ means $\mu(\mathbf{v}, \alpha)$ and η means $\eta(\mathbf{v},\alpha)$, the arguments having been dropped for convenience and to emphasize that $\mu(\mathbf{v},\alpha)$ and $\eta(\mathbf{v},\alpha)$ are constant over each set D_i. Then, we have

$$J(\mathbf{v}, \alpha) = \sum_{l=0}^{\xi} (\tau_{2l+3-\eta} - \tau_{2l+2-\eta}). \tag{6.52}$$

Let $k = 2l + 2 - \eta$. Then we have

$$J(\mathbf{v},\alpha) = \sum_{k=2-\eta}^{2\xi+2-\eta} \tau_k(-1)^{(k+\eta-1)}. \tag{6.53}$$

From Equation 6.27, we get

$$\frac{\partial \tau_j}{\partial \alpha} = \begin{cases} 0, & j = 0, \mu + 1, \\ \dfrac{-\langle \mathbf{v}, e^{\mathbf{A}\tau_j}\mathbf{b}\rangle}{\langle \alpha\mathbf{v}, \mathbf{A}e^{\mathbf{A}\tau_j}\mathbf{b}\rangle}, & j = 1, 2, \ldots, \mu, \end{cases} \tag{6.54}$$

$$\frac{\partial \tau_j}{\partial \theta_i} = \begin{cases} 0, & j = 0, \mu + 1,^2 \\ \dfrac{-\langle \alpha\,\mathbf{i}, e^{\mathbf{A}\tau_j}\mathbf{b}\rangle}{\langle \alpha\mathbf{v}. \mathbf{A}e^{\mathbf{A}\tau_j}\mathbf{b}\rangle} \dfrac{\partial \mathbf{h}(\theta)}{\partial \theta_i} & \begin{array}{l} i = 1, 2, \ldots, n, \\ j = 1, 2, \ldots, \mu + 1. \end{array} \end{cases} \tag{6.55}$$

Here $\mathbf{h}(\boldsymbol{\theta})$ is the hyperspherical coordinate transformation of Definition 4.11 which is used to restrict the derivative to the set D (that is, $\mathbf{v} \in \partial S$). Clearly, the right-hand sides of Equation 6.54 and 6.55 have continuous partial derivatives of all orders if $(\mathbf{v}, \alpha) \neq Q$. It follows that $J(\mathbf{v}, \alpha)$ in Equation 6.53 has continuous partial derivatives of all orders on any D_i.

From Definition 6.11 and Equations 6.17, 6.45, 6.50, 6.49, and 6.51, we find that

$$\mathbf{M}(v,\alpha) = e^{-\mathbf{A}T} \left\{ \mathbf{v} + \sum_{l=0}^{\xi} \omega_{2l+2-\eta} \int_{\tau_{2l+2-\eta}}^{\tau_{2l+3-\eta}} e^{\mathbf{A}s}\mathbf{b}\, ds \right\}. \tag{6.56}$$

[2] Here $\mathbf{i}$ is a column vector with all zero elements except the ith element which is unity.

Again, let $k = 2l + 2 - \eta$, and since $-\mathbf{A}$ is assumed positive definite, we may write

$$\mathbf{M}(\mathbf{v},\alpha) = e^{-\mathbf{A}T}\left\{ v + \mathbf{A}^{-1} \sum_{k=2-\eta}^{2\xi+2-\eta} \omega_k e^{\mathbf{A}\tau_k}\mathbf{b}(-1)^{(k+\eta-1)}\right\}. \tag{6.57}$$

From Equations 6.46, 6.47, 6.54, and 6.55, we conclude that Equation 6.57 has continuous partial derivatives to all order.

Q.E.D.

It follows that both $\mathbf{M}(\mathbf{v},\alpha)$ and $J(\mathbf{v},\alpha)$ are differentiable to all orders (on each set D_i) in the sense of Frechet (see Appendix B and [15] p. 90). Let $\mathbf{M}_\mathbf{v}(\mathbf{v},\alpha)$, $\mathbf{M}_\alpha(\mathbf{v},\alpha)$, $J_\mathbf{v}(\mathbf{v},\alpha)$, and $J_\alpha(\mathbf{v},\alpha)$ denote these respective first derivatives. That is, we have

$$\mathbf{M}_\mathbf{v}(\mathbf{v},\alpha) = \left[\frac{\partial \mathbf{M}(\mathbf{v},\alpha)}{\partial v_1} \quad \frac{\partial \mathbf{M}(\mathbf{v},\alpha)}{\partial v_2} \quad \cdots \quad \frac{\partial \mathbf{M}(\mathbf{v},\alpha)}{\partial v_n} \right], \tag{6.58}$$

$$J_\mathbf{v}(\mathbf{v},\alpha) = \begin{bmatrix} \dfrac{\partial J(\mathbf{v},\alpha)}{\partial v_1} \\ \vdots \\ \dfrac{\partial J(\mathbf{v},\alpha)}{\partial v_n} \end{bmatrix}. \tag{6.59}$$

From Equations 6.53 and 6.54, we have

$$J_\alpha(\mathbf{v},\alpha) = \sum_{k=1}^{\mu} (-1)^{k+\eta} \frac{\langle v, e^{\mathbf{A}\tau_k}\mathbf{b}\rangle}{\langle \alpha\mathbf{v}, \mathbf{A}e^{\mathbf{A}\tau_k}\mathbf{b}\rangle}, \qquad (\mathbf{v},\alpha) \notin Q. \tag{6.60}$$

Examination of the structure of $\operatorname{dez}\{\langle \alpha\mathbf{v}, e^{\mathbf{A}\tau}\mathbf{b}\rangle\}$ (see Figure 6.2) will show that for $(\mathbf{v},\alpha) \notin Q$,

$$\operatorname{sgn}\{\langle \alpha\mathbf{v}, \mathbf{A}e^{\mathbf{A}\tau_k}\mathbf{b}\rangle\} = (-1)^{k+\eta} \operatorname{sgn}\{\langle \alpha\mathbf{v}, e^{\mathbf{A}\tau_k}\mathbf{b}\rangle\}. \tag{6.61}$$

And, since $|\langle \mathbf{v}, e^{\mathbf{A}\tau_k}\mathbf{b}\rangle| = 1, k = 1, 2, \ldots, \mu$, we have

$$J_\alpha(\mathbf{v},\alpha) = \frac{1}{\alpha} \sum_{k=1}^{\mu} \frac{1}{|\langle \alpha\mathbf{v}, \mathbf{A}e^{\mathbf{A}\tau_j}\mathbf{b}\rangle|}. \tag{6.62}$$

In words, $J_\alpha(\mathbf{v},\alpha)$ is strictly positive wherever it exists, which is to be expected in view of Theorem 6.2.

Definition 6.12

Let $(\mathbf{v}, \alpha) \in D_i$ (for some i) and let $\theta = (1/r)h^{-1}(v)$, then $G(\mathbf{v}, \alpha)$ is the $n \times n$ matrix defined as

$$\mathbf{G}(\mathbf{v},\alpha) = \left[r(\mathbf{M_v}(v,\alpha) - \frac{1}{J_\alpha(\mathbf{v},\alpha)} (\mathbf{M}_\alpha(\mathbf{v},\alpha)\rangle \langle J_\mathbf{v}(v,\alpha))\, \mathbf{h}_\theta(\theta) \right. \vdots$$

$$\left. \times \frac{1}{J_\alpha(\mathbf{v},\alpha)} \mathbf{M}_\alpha(\mathbf{v},\alpha) \right]. \tag{6.63}$$

The motivation for $\mathbf{G}(\mathbf{v}, \alpha)$ is that it describes the first-order effects, on the solution of the initial-value problem of Equation 6.15 and 6.16, of changing $\mathbf{v}$ and J. That is, if we have

$$J_{k+1} = J_k + \Delta J_k, \tag{6.64}$$

and

$$\mathbf{v}_{k+1} = \mathbf{v}_k + \Delta \mathbf{v}_k, \tag{6.65}$$

and

$$\alpha_{k+1} = \alpha_k + \Delta\alpha_k. \tag{6.66}$$

then, to first-order terms, the changes $\Delta J_k, \Delta \mathbf{v}_k, \Delta\alpha_k$, and

$$\Delta \mathbf{x}_k = \mathbf{x}_{k+1} - \mathbf{x}_k \tag{6.67}$$

are related by the approximations

$$J_{k+1}(\mathbf{v}_{k+1}, \alpha_{k+1}) \cong J_k(\mathbf{v}_k, \alpha_k) + \langle J_\mathbf{v}(\mathbf{v}_k, \alpha_k), \Delta \mathbf{v}_k \rangle$$

$$+ J_\alpha(\mathbf{v}_k, \alpha_k)\, \Delta\alpha_k, \tag{6.68}$$

$$\Delta \mathbf{x}_k = \mathbf{M}(\mathbf{v}_{k+1}, \alpha_{k+1}) - \mathbf{M}(\mathbf{v}_k, \alpha_k), \tag{6.69}$$

from whence we have

$$\Delta \mathbf{x}_k \cong \mathbf{M}_\mathbf{v}(\mathbf{v}_k, \alpha_k)\Delta \mathbf{v}_k + \mathbf{M}_\alpha(\mathbf{v}_k, \alpha_k)\Delta\alpha_k. \tag{6.70}$$

From Equations 6.67 and 6.68, we have

$$\Delta \mathbf{x}_k \cong \left[\mathbf{M_v}(\mathbf{v}_k, \alpha_k) - \frac{1}{J_\alpha(\mathbf{v}_k, \alpha_k)} \mathbf{M}_\alpha(\mathbf{v}_k, \alpha_k)\rangle \langle J_\mathbf{v}(\mathbf{v}_k, \alpha_k) \right]$$

$$\times \Delta \mathbf{v}_k + \frac{1}{J_\alpha(\mathbf{v}_k,\alpha_k)} \mathbf{M}_\alpha(\mathbf{v}_k, \alpha_k)\Delta J_k, \tag{6.71}$$

using

$$\Delta \mathbf{v}_k = r h_\theta(\theta_k)\, \Delta \theta_k, \tag{6.72}$$

where

$$\theta_k = \frac{1}{r}\, \mathbf{h}^{-1}(\mathbf{v}_k), \tag{6.73}$$

$$\Delta \mathbf{x}_k \cong \mathbf{G}(\mathbf{v}, \alpha_k) \begin{bmatrix} \Delta \theta_k \\ \hdashline \Delta J_k \end{bmatrix}. \tag{6.74}$$

Theorem 6.4

The $n \times n$ matrix $\mathbf{G}(\mathbf{v}, \alpha)$ of Definition 6.11 has rank n for $(\mathbf{v}, \alpha)$ on D wherever it exists (that is, on D/Q).

Proof:

From Equations 6.54 and 6.57, we have

$$\mathbf{M}_\alpha(\mathbf{v}, \alpha) = e^{\mathbf{A}T} \left\{ \sum_{k=1}^{\mu} \omega_k e^{\mathbf{A}\tau_k} \mathbf{b} (-1)^{(k+\eta)} \frac{\langle \mathbf{v}, e^{\mathbf{A}\tau_k}\mathbf{b} \rangle}{\langle \alpha \mathbf{v}, \mathbf{A} e^{\mathbf{A}\tau_k}\mathbf{b} \rangle} \right\}, \tag{6.75}$$

and using Equation 6.62, we get

$$\mathbf{M}_\alpha(\mathbf{v}, \alpha) = e^{-\mathbf{A}T} \left\{ \sum_{k=1}^{\mu} \frac{\omega_k}{\alpha\, |\langle \alpha \mathbf{v}, \mathbf{A} e^{\mathbf{A}\tau_k} \mathbf{b} \rangle|} e^{\mathbf{A}\tau_k} \mathbf{b} \right\}. \tag{6.76}$$

From Equations 6.55 and 6.57, we obtain

$$\mathbf{M}_\mathbf{v}(\mathbf{v}, \alpha) = e^{-\mathbf{A}T} \left\{ \mathbf{I} + \sum_{k=1}^{\mu} \frac{\alpha \boldsymbol{\omega}_k (-1)^{(k+\eta)} e^{\mathbf{A}\tau_k} \mathbf{b} \rangle}{\langle \alpha \mathbf{v}, \mathbf{A} e^{\mathbf{A}\tau_k}\mathbf{b} \rangle} \langle e^{\mathbf{A}\tau_k} \mathbf{b} \right\}, \tag{6.77}$$

and using Equations 6.55 and 6.61, we find that

$$\mathbf{M}_\mathbf{v}(\mathbf{v}, \alpha) = e^{-\mathbf{A}T} \left\{ \mathbf{I} + \sum_{k=1}^{\mu} \frac{\alpha}{|\langle \alpha \mathbf{v}, \mathbf{A} e^{\mathbf{A}\tau_k} \mathbf{b} \rangle|} e^{\mathbf{A}\tau_k} \mathbf{b} \rangle \langle e^{\mathbf{A}\tau_k} \mathbf{b} \right\}. \tag{6.78}$$

From Equations 6.53 and 6.55, we have

$$J_\mathbf{v}(\mathbf{v}, \alpha) = \sum_{k=1}^{\mu} \frac{(-1)^{(k+\eta)}\alpha}{\langle \alpha \mathbf{v}, \mathbf{A} e^{\mathbf{A}\tau_k} \mathbf{b} \rangle} e^{\mathbf{A}\tau_k} \mathbf{b}, \tag{6.79}$$

and using Equations 6.41 and 6.51, we obtain

$$J_{\mathbf{v}}(\mathbf{v}, \alpha) = \sum_{k=1}^{\mu} \frac{\alpha\omega_k}{|\langle \alpha\mathbf{v}, \mathbf{A}e^{\mathbf{A}\tau_k}\mathbf{b}\rangle|} e^{\mathbf{A}\tau_k}\mathbf{b}. \tag{6.80}$$

Hence, we have

$$\mathbf{G}(\mathbf{v},\alpha) = \left\{ re^{-\mathbf{A}\mathrm{T}}[\mathbf{I} + \mathbf{E}]\mathbf{h}_\theta(\theta) \,\vdots\, \frac{1}{J_\alpha(\mathbf{v},\alpha)} \mathbf{M}_\alpha(\mathbf{v},\alpha) \right\}, \tag{6.81}$$

where, using Equations 6.62, 6.63, 6.76, 6.78, and 6.79, and letting

$$\rho_k = \frac{1}{|\langle \alpha\mathbf{v}, \mathbf{A}e^{\mathbf{A}\tau_k}\mathbf{b}\rangle|}, \quad k = 1, 2, \ldots, \mu, \tag{6.82}$$

and

$$\mathbf{z}_k = e^{\mathbf{A}\tau_k}\mathbf{b}, \tag{6.83}$$

we obtain

$$\mathbf{E} = \alpha \left\{ \left(\sum_{k=1}^{\mu} \rho_k \mathbf{z}_k \rangle \langle \mathbf{z}_k \right) - \frac{1}{\left(\sum_{k=1}^{\mu} \rho_k \right)} \left(\sum_{k=1}^{\mu} \rho_k \omega_k \mathbf{z}_k \right) \rangle \langle \left(\sum_{k=1}^{\mu} \rho_k \omega_k \mathbf{z}_k \right) \right\}. \tag{6.84}$$

Following the proof of Theorem 4.5, one need only show that $\mathbf{E}$ is positive semidefinite and then find a vector which is both linearly dependent on the nth column of $\mathbf{G}(\mathbf{v},\alpha)$ and orthogonal to the other $n - 1$. Since

$$\langle \alpha\mathbf{v}, \mathbf{z}_k \rangle = \omega_k, \tag{6.85}$$

it follows that

$$\langle \alpha\mathbf{v}, \mathbf{E}\,\alpha\mathbf{v} \rangle = 0. \tag{6.86}$$

Hence, $\mathbf{E}$ is not definite. The quadratic form satisfies

$$\langle \mathbf{x}, \mathbf{E}\mathbf{x} \rangle \geqslant 0 \tag{6.87}$$

if and only if

$$\left(\sum_{k=1}^{\mu} \rho_k \right) \left(\sum_{k=1}^{\mu} \rho_k \langle \mathbf{x}, \mathbf{z}_k \rangle^2 \right) \geqslant \left(\sum_{k=1}^{\mu} \rho_k \omega_k \langle \mathbf{x}, \mathbf{z}_k \rangle \right)^2. \tag{6.88}$$

Since $\omega_k{}^2 = 1, k = 1, 2, \ldots, \mu$ (see Equation 6.42), Equation 6.88 becomes

$$\sum_{k=1}^{\mu} \sum_{i=1}^{\mu} \rho_i \rho_k \{ c_k{}^2 - c_k c_i \}, \tag{6.89}$$

where

$$c_i \equiv \langle \mathbf{x}, \mathbf{z}_i \rangle \, \omega_k. \tag{6.90}$$

In Equation 6.89 the terms for $i = k$ cancel. When we do this and rearrange, we get

$$\sum_{k=i+1}^{\mu} \sum_{i=1}^{\mu} \rho_i \rho_k \{ c_i{}^2 + c_k{}^2 - 2c_i c_k \} \geqslant 0, \tag{6.91}$$

or

$$\sum_{k=i+1}^{\mu} \sum_{i=1}^{\mu-1} \rho_i \rho_k \{ (c_i - c_k)^2 \} \geqslant 0. \tag{6.92}$$

But Equation 6.92 holds term by term, since $\rho_i > 0$. Hence, $\mathbf{E}$ is positive semidefinite. From Property 2 of $\mathbf{h}_\theta(\theta)$ and Equation 6.87, the vector $\alpha e^{\mathbf{A}'T}\mathbf{v}$ (the costate vector at $\tau = T$) is orthogonal to the first $n - 1$ colums of $\mathbf{G}$. From Equations 6.63 and 6.97, we have

$$\langle \alpha \, e^{\mathbf{A}'T}\mathbf{v}, \frac{1}{J_\alpha(\mathbf{v}, \alpha)} \, M_\alpha(\mathbf{v}, \alpha) \rangle = 1, \tag{6.93}$$

and $\mathbf{G}(\mathbf{v}, \alpha)$ has rank n. Q.E.D.

Just as in the case of Problem Z_1, the addition of the matrix $\mathbf{I}$ to E in Equation 6.82 represents the effect on $\mathbf{G}(\mathbf{v}, \alpha)$ of replacing the origin with the hypersphere as a target. Since $\mathbf{E}$ is positive semidefinite, this addition (that is, modifying the problem) is sufficient, precisely in the same manner as in Problem Z_1, to ensure that $\mathbf{G}(\mathbf{v}, \alpha)$ has rank n wherever it exists. Once again, the matrices $\mathbf{G}(\mathbf{v}, \alpha)$ and $\mathbf{G}^{-1}(\mathbf{v}, \alpha)$ are regarded as piecewise continuous functions from D into $R_n \times R_n$. Obviously, the discontinuities occur on Q.

6.4 The Iterative Procedure

Again the iterative procedure is based on the fact that $\mathbf{G}(\mathbf{v}, \alpha)$ and $\mathbf{G}^{-1}(\mathbf{v}, \alpha)$ exist almost everywhere on D. In the following, it is assumed that (1) $T > T^*$; (2) $\|e^{\mathbf{A}T}\mathbf{x}_0\| > r$ (that is, assuming that one has already tested for $u^*_{(0,T]} = 0$), and (3) that the sequence of guesses $(\mathbf{v}_k, \alpha_k)$ remain in D/Q. Suppose that the kth guess $(\mathbf{v}_k, \alpha_k)$ has been made. The following recursion formulas describe the iterative procedure:

$$\mathbf{x}_k = \mathbf{M}(\mathbf{v}_k, \alpha), \tag{6.94}$$

$$\mathbf{e}_k = \mathbf{x}_0 - \mathbf{x}_k, \tag{6.95}$$

where

$$\mathbf{x}_0 = \mathbf{M}(\mathbf{v}^*, \alpha^*). \tag{6.96}$$

Let

$$\theta_k = \frac{1}{r}\, \mathbf{h}^{-1}(\mathbf{v}_k), \tag{6.97}$$

and

$$\mathbf{G}_k = \left[r\left\{\mathbf{M}_\mathbf{v}(\mathbf{v}_k, \alpha_k) - \frac{1}{J_\alpha(\mathbf{v}_k, \alpha_k)}\, \mathbf{M}_\alpha(\mathbf{v}_k, \alpha_k)\rangle \langle \mathbf{J}_\mathbf{v}, \alpha_k\rangle\right\} \times \mathbf{h}_\theta(\theta) \;\vdots\; \frac{1}{J_\alpha(\mathbf{v}_k, \alpha_k)}\, \mathbf{M}_\alpha(\mathbf{v}_k, \alpha_k) \right], \tag{6.98}$$

Also let

$$\begin{bmatrix} \Delta\hat{\theta}_k \\ \text{----} \\ \Delta\hat{J}_k \end{bmatrix} = G_k^{-1}\mathbf{e}_k, \tag{6.99}$$

$$\Delta\hat{\mathbf{y}}_k = \mathbf{h}_\theta(\theta_k)\,\Delta\hat{\theta}_k, \tag{6.100}$$

$$\Delta\hat{\alpha}_k = \frac{1}{J_\alpha(\mathbf{v}_k, \alpha_k)}\,(\Delta\hat{J}_k - \langle \mathbf{J}_\mathbf{v}(\mathbf{v}_k, \alpha_k), \Delta\hat{\mathbf{y}}_k\rangle), \tag{6.101}$$

$$\mathbf{v}_{k+1}(\gamma_k) = \frac{r(\mathbf{v}_k + \gamma_k\Delta\hat{\mathbf{y}}_k)}{\|\mathbf{v}_k + \gamma_k\Delta\hat{\mathbf{y}}_k\|}, \tag{6.102}$$

and

$$\alpha_{k+1}(\gamma_k) = \alpha_k + \gamma_k\Delta\hat{\alpha}_k. \tag{6.103}$$

Choose γ_k^* in $(0, 1]$ such that

$$\|\mathbf{e}_{k+1}\| = \min_{\gamma_k \in (0,1]} \|\mathbf{x}_0 - \mathbf{M}(\mathbf{v}_{k+1}(\gamma_k), \alpha_{k+1}(\gamma_k))\|, \tag{6.104}$$

and then choose

$$\mathbf{v}_{k+1} = \mathbf{v}_{k+1}(\gamma_k^*), \tag{6.105}$$

$$\alpha_{k+1} = \alpha_{k+1}(\gamma_k^*). \tag{6.106}$$

The role of $\mathbf{h}(\theta)$ and $\Delta\hat{\mathbf{y}}_k$ is the same here as in the iterative procedure of Section 4.4. The difference is in the role played by $J(\mathbf{v}, \alpha)$. This function is first used to find the matrix $\mathbf{G}$ in the form of Equation 6.99, which has been shown in Theorem 6.4 to have rank n. The conversion back to variables $(\mathbf{v}, \alpha)$ is made using Equation 6.102. The iterative procedure could have been set up in terms of the variables $\mathbf{v}$ and J instead of $\mathbf{v}$ and α. However, there is good reason for choosing α instead of J. Given α and $\mathbf{v}$, it is easy numerically to find J, but given J and $\mathbf{v}$, the problem of finding α is numerically much more difficult in terms of the amount of computation required.

6.5 Convergence

All of the conclusions and comments of Section 4.5 apply to Problem Z_3 if T is replaced by α. Rather than repeat that entire section here, Lemma 4.4 is proved for Problem Z_3. The remainder of Section 4.5 applies similarly, with appropriate slight changes in notation.

Lemma 6.5

If the vector $\Delta\hat{\mathbf{y}}_k$ and the scalar $\Delta\hat{T}_k$ are such that the points $(\mathbf{v}_{k+1}(\gamma_k), \alpha_{k+1}(\gamma_k))$ defined by Equations 6.102 and 6.103 are in the same subset D_i as the point $(\mathbf{v}_k, \alpha_k)$ (the current guess) for γ_k in $[0, \delta_k], \delta_k > 0$, then there exists a positive real number N_k, such that the function $\|\mathbf{e}_{k+1}(\gamma_k)\|$ is bounded on $[0, \delta_k]$ by the relation

$$\|\mathbf{e}_{k+1}(\gamma_k)\| \leq |1 - \gamma_k| \, \|\mathbf{e}_k\| + \gamma_k^2 N_k, \tag{6.107}$$

where $\mathbf{e}_k$ is defined by Equation 6.96.

Proof:

Using the chain rule, the fact that $\mathbf{M}(\mathbf{v}, \alpha)$ is twice differentiable on D/Q, and the mean-value theorem, and letting

$$\mathbf{M}(\gamma_k) = \mathbf{M}(\mathbf{v}_{k+1}(\gamma_k), \alpha_{k+1}(\gamma_k)), \tag{6.109}$$

we have

$$\mathbf{M}(\gamma_k) = \mathbf{M}(0) + \mathbf{M_v}(k)\left.\frac{d\mathbf{v}_{k+1}}{d\gamma_k}(\gamma_k)\right|_{\gamma_k=0}\gamma_k + \mathbf{M}_\alpha(k)\left.\frac{d\alpha_{k+1}}{d\gamma_k}(\gamma_k)\right|_{\gamma_k=0}\gamma_k$$

$$+\frac{1}{2}\gamma_k{}^2 M^{(2)}, \tag{6.109}$$

where $M^{(2)}$ is the second-derivative term evaluated somewhere on $[0, \delta_k]$. But,

$$\mathbf{e}_{k+1} = \mathbf{x}_0 - \mathbf{M}(\gamma_k), \tag{6.110}$$

$$\mathbf{e}_k = \mathbf{x}_0 - \mathbf{M}(0), \tag{6.111}$$

$$\left.\frac{d\alpha_{k+1}}{d\gamma_k}(\gamma_k)\right|_{\gamma_k=0} = \Delta\hat{\alpha}_k, \tag{6.112}$$

$$\left.\frac{d\mathbf{v}_{k+1}}{d\gamma_k}(\gamma_k)\right|_{\gamma_k=0} = \Delta\hat{\mathbf{y}}_k, \tag{6.113}$$

whence, we have

$$\mathbf{e}_{k+1}(\gamma_k) = \mathbf{e}_k - \mathbf{M_v}(k)\Delta\mathbf{y}_k\gamma_k - \mathbf{M}_\alpha(k)\Delta\alpha_k\gamma_k - \tfrac{1}{2}\gamma_k{}^2\mathbf{M}^{(2)}. \tag{6.114}$$

Then, using Equations 6. 98, 6. 99, 6. 100, 6 101, and the triangle inequality for norms, we have

$$N_k = \tfrac{1}{2}\,\|M^{(2)}\|\,. \tag{6.115}$$

Q.E.D.

Because of the strong similarity between Problems Z_1 and Z_3 in terms of the set Q and the convergence arguments, it is to be expected that the iterative procedure will behave similarly in both cases.

6.6 The Initial Guess

The method of problem separation described in Chapter 4 has little appeal unless one can solve the lower-order problems rapidly to find estimates for J^*. This is not possible in Problem Z_3 when the eigenvalues are complex. Two other initial-guess procedures are proposed here.

First Procedure. Suppose that α^* and $\mathbf{v}^*$ are known. Then, we have

$$\langle \mathbf{v}^*, e^{\mathbf{A}T}\mathbf{x}_0\rangle = \langle \mathbf{v}^*, \mathbf{v}^*\rangle + \frac{1}{\alpha^*}\int_0^T \langle \alpha^*(\mathbf{v}^*, e^{\mathbf{A}T}\mathbf{b}\rangle \times \text{dez}\{\alpha^*\mathbf{v}^*, e^{\mathbf{A}s}\mathbf{b}\rangle\}\, ds, \tag{6.116}$$

that is,

$$\langle \mathbf{v}^*, e^{\mathbf{A}T}\mathbf{x}_0\rangle > r^2. \tag{6.117}$$

Inequality 6. 117 suggests that a good first guess would be

$$\mathbf{v}_1 = \frac{r e^{\mathbf{A}T}\mathbf{x}_0}{\|e^{\mathbf{A}T}\mathbf{x}_0\|}. \tag{6.118}$$

Such a guess would satisfy Inequality 6. 117 and, furthermore,

$$\langle \mathbf{v}_1, e^{\mathbf{A}T}\mathbf{x}_0\rangle = r\,\|e^{\mathbf{A}T}\mathbf{x}_0\| \tag{6.119}$$

would be increasing with $\|e^{\mathbf{A}T}\mathbf{x}_0\|$, which is desirable, since as T approaches T*, $\|e^{\mathbf{A}T}\mathbf{x}_0\|$ increases, and the function $\text{dez}\{\langle \alpha^*, \mathbf{v}^*, e^{\mathbf{A}s}\mathbf{b}\rangle\}$ approaches the function $\text{sgn}\{(\mathbf{v}^*, e^{\mathbf{A}s}\mathbf{b}\rangle\}$ that maximizes the scalar product $\langle \mathbf{v}^*, e^{\mathbf{A}T}\mathbf{x}_0\rangle$. It remains to choose α_1. Consider the function

$$f(\mathbf{x}_0; \alpha) \equiv \langle \mathbf{v}_1, e^{\mathbf{A}T}(\mathbf{x}_0 - \mathbf{M}(\mathbf{v}_1, \alpha))\rangle, \tag{6.120}$$

where $\mathbf{v}_1$ is given by Equation 6. 119. Since

$$\mathbf{M}(\mathbf{v}_1, 0) = e^{-\mathbf{A}T}\mathbf{v}_1, \tag{6.121}$$

we have

$$f(\mathbf{x}_0; 0) = r\|e^{\mathbf{A}T}\mathbf{x}_0\| - r^2 > 0 \tag{6.122}$$

Also, $f(\mathbf{x}_0; \alpha)$ is continuous in α (since $\mathbf{M}(v,\alpha)$ is continuous), and if $(\mathbf{v}_1, \alpha) \in Q$,

$$\frac{df}{d\alpha}(\mathbf{x}_0; \alpha) = -\langle \mathbf{v}_1, e^{\mathbf{A}T}\mathbf{M}_\alpha(\mathbf{v}_1, \alpha)\rangle, \tag{6.123}$$

which from Equations 6. 41 and 6. 76 is strictly negative. Now, $e^{\mathbf{A}'T}\mathbf{v}_1$ has the direction of the costate vector that generated the fuel-optimal trajectory from $\mathbf{M}(\mathbf{v}_1, \alpha)$. Then $f(\mathbf{x}_0; \alpha)$ is the scalar product of the error $\mathbf{x}_0 - \mathbf{M}(v, \alpha)$ with a vector whose direction is that of the costate vector. As is well known, the costate vector is normal to a support hyperplane of the minimum-cost surface at $\mathbf{M}(\mathbf{v}_1, \alpha)$. Comparing Equations 6. 17 and 6. 26 and noting that S is convex, we conclude that the minimum-fuel surfaces are convex. It follows that if

$T > T^*$, there exists an $\alpha_1 > 0$ such that $f(\mathbf{x}_0, \alpha_1) = 0$ (that is, the error $\mathbf{x}_0 - \mathbf{M}(\mathbf{v}_1, \alpha_1)$ lies in the support hyperplane of the minimum-fuel surface passing through $\mathbf{M}(\mathbf{v}_1, \alpha_1)$). It is proposed to choose $\mathbf{v}_1$ in accordance with Equation 6.136 and then choose α_1 such that $f(\mathbf{x}_0; \alpha_1) = 0$.

Second Procedure. Again choose

$$\mathbf{v}_1 = \frac{r e^{\mathbf{A}T}\mathbf{x}_0}{\| e^{\mathbf{A}T}\mathbf{x}_0 \|} . \tag{6.124}$$

However, instead of finding α_1 such that $f(\mathbf{x}_0; \alpha_1) = 0$, find the α_1 which minimizes $\|\mathbf{x}_0 - \mathbf{M}(\mathbf{v}_1, \alpha)\|$, that is, carry out a search by increasing α in steps until the first minimum of $\|\mathbf{x}_0 - \mathbf{M}(\mathbf{v}_1, \alpha)\|$ is found). This search can be carried out in essentially the same manner as for finding a zero of $f(\mathbf{x}_0; \alpha)$, because, if $\|\mathbf{x}_0 - \mathbf{M}(\mathbf{v}_1, \alpha)\|$ is a minimum at $\alpha = \hat{\alpha}$, then

$$\langle \mathbf{x}_0 - \mathbf{M}(\mathbf{v}_1, \hat{\alpha}), \mathbf{M}_\alpha(\mathbf{v}_1, \hat{\alpha})\rangle = 0. \tag{6.125}$$

6.7 The Multi-input Case

Very briefly, the generalization to the multi-input case can be carried out by summing over the m controllers exactly as was done in Section 4.7. For example,

$$\mathbf{M}(\mathbf{v}, \alpha) = e^{-\mathbf{A}T} \left\{ \mathbf{v} + \sum_{i=1}^{m} \int_0^T e^{\mathbf{A}s}\mathbf{b}_i \operatorname{dez}\{\langle \alpha\mathbf{v}, e^{\mathbf{A}s}\mathbf{b}_i\rangle\} \, ds \right\}, \tag{6.126}$$

and

$$\mathbf{G} = r e^{-\mathbf{A}T} \left\{ \mathbf{I} + \sum_{i=1}^{m} \mathbf{E}_i \right\} \mathbf{h}_\theta(\theta)\mathbf{M}_\alpha . \tag{6.127}$$

See Equations 4.125, 4.137, 6.17, and 6.84. Note as before (Section 4.7) that one must consider m functions $\mu_i(\mathbf{v}, \alpha)$, $i = 1, 2, \ldots, m$; m functions $\eta_i(\mathbf{v}, \alpha)$, $i = 1, 2, \ldots, m$; m sets of switch times $\tau_{ji}(\mathbf{v}, \alpha)$, $u = 1, 2, \ldots, \mu_i$, $i = 1, 2, \ldots, m$; and m sets of functions $\omega_{ji}(\mathbf{v}, \alpha)$, $j = 1, 2, \ldots, \mu_i$, $i = 1, 2, \ldots, m$. However, all definitions extend trivially as in Section 4.7. The only involved part of the extension to m inputs is the proof that the equivalent of Inequality 6.88 holds. If the extension is carried out, Inequality 6.88 becomes[3]

$$\left(\sum_{i=1}^{m}\sum_{k=1}^{\mu_i} \rho_{ki}\right)\left(\sum_{i=1}^{m}\sum_{k=1}^{\mu_i} \rho_{ki}\omega_{ki}^2\langle \mathbf{x}, \mathbf{z}_{ki}\rangle^2\right) \geqslant \left(\sum_{i=1}^{m}\sum_{k=1}^{\mu_i} \rho_{ki}\omega_{ki}\langle \mathbf{x}, \mathbf{z}_{ki}\rangle^2\right)^2 \tag{6.128}$$

[3] Here $\omega_{ki}^2 = 1$, $k = 1, 2, \ldots, \mu_i$, $i = 1, 2, m$.

(see Equations 6. 82, 6. 83, and 6. 88).

From Theorem 6. 4, we have

$$\left(\sum_{k=1}^{\mu_i} \rho_{ki}\right)\left(\sum_{k=1}^{\mu_i} \rho_{ki}\omega_{ki}^2\langle \mathbf{x}, \mathbf{z}_k\rangle^2\right) \geqslant \left(\sum_{k=1}^{\mu_i} \rho_{ki}\omega_{ki}\langle \mathbf{x}, \mathbf{z}_{ki}\rangle\right)^2$$

$$\text{for } i = 1, 2, \ldots, m. \tag{6.129}$$

It follows that Relation 6. 128 holds if

$$\left(\sum_{k=1}^{\mu_l} \rho_{kl}\right)\left(\sum_{j=1}^{\mu_n} \rho_{jn}\omega_{jn}^2\langle \mathbf{x}, \mathbf{z}_{jn}\rangle^2\right) + \left(\sum_{j=1}^{\mu_n} \rho_{jn}\right)\left(\sum_{k=1}^{\mu_l} \rho_{kl}\omega_{kl}^2\langle \mathbf{x}, \mathbf{z}_{kl}\rangle^2\right)$$

$$\geqslant 2 \sum_{k=1}^{\mu_l}\sum_{j=1}^{\mu_n} \rho_{jn}\rho_{kl}\omega_{jn}\omega_{kl}\langle \mathbf{x}, \mathbf{z}_{jl}\rangle\langle \mathbf{x}, \mathbf{z}_{kl}\rangle, \tag{6.130}$$

for $l = 1, 2, \ldots, m$, and $n = 1, 2, \ldots, m$, and $n \neq l$.

But Relation 6. 130 may be rewritten

$$\sum_{k=1}^{\mu_l}\sum_{j=1}^{\mu_n} \rho_{kl}\rho_{jn}(\omega_{jn}\langle \mathbf{x}, \mathbf{z}_{jn}\rangle - \omega_{kl}\langle \mathbf{x}, \mathbf{z}_{kl}\rangle)^2 \geqslant 0 \tag{6.131}$$

which is obviously true, since $\rho_{kl} > 0$, $k = 1, 2, \ldots, \mu_l$, $l = 1, 2 \ldots, m$. Hence, the iterative procedure may be applied to the multi-input case as well as to the time-varying case (see Section 4. 8).

7. The Minimum-Effort Control Problem

7.1 Introduction

Problem Z_4 is considered in this chapter (Definition 1.14). Because of the many similarities between Problems Z_3 and Z_4, Chapter 6 is referred to for a proof or a definition wherever possible. The problem is briefly restated in Section 7.2, and the function $\mathbf{M}(\mathbf{v},\alpha)$ is defined. In Section 7.3, it is established that $\mathbf{M}(\mathbf{v},\alpha)$ has continuous first partial derivatives everywhere, and that the $n \times n$ matrix $\mathbf{G}(\mathbf{v},\alpha)$ defined on the set D, as in Chapter 6, has rank n. In the remainder of the chapter, it is shown that the iterative procedure, convergence, and the initial-guess procedures, as given in Chapter 6, apply here.

7.2 The Function M

Briefly restating Problem Z_4 (Definition 1.14) we have the following. Given

$$\dot{\mathbf{x}}(t) = \mathbf{A}\mathbf{x}(t) + \mathbf{b}u(t), \qquad \mathbf{x}(0) = \mathbf{x}_0, \tag{7.1}$$

$$S = \{\mathbf{x}: \langle \mathbf{x}, \mathbf{x} \rangle - r^2 \leq 0\}, \qquad r > 0, \tag{7.2}$$

$$J = \int_0^T \frac{1}{\rho} |u(s)|^\rho \, ds, \qquad |u(s)| \leq 1, \rho > 1, \tag{7.3}$$

$$T \text{ is fixed}, \qquad T > T^*, \tag{7.4}$$

$$\mathbf{x}_0 \in C/S. \tag{7.5}$$

Find the control that steers the state from an initial condition $\mathbf{x}_0 \in C/S$ (Definition 1.6) to an unspecified terminal state $\mathbf{v} \in S$, such that the elapsed time is T and the cost functional J is minimized. Applying the Minimum Principle, one concludes that if the optimal terminal state $\mathbf{v}^* \in \partial S$, then NC10 of Theorem 1.2 applies, and

$$\mathbf{p}^*(T) = \alpha^* \mathbf{v}^*, \qquad \alpha > 0, \tag{7.6}$$

And, if $\mathbf{v}^* \in S/\partial S$, then NC11 applies, and

$$\mathbf{p}^*(T) = \mathbf{0}. \tag{7.7}$$

The Hamiltonian for Problem Z_3 is

$$H = p_0 \frac{1}{\rho} |u|^\rho + \langle \mathbf{p}(t), \mathbf{A}\mathbf{x}(t)\rangle + \langle \mathbf{b}, \mathbf{p}(t)\rangle u. \tag{7.8}$$

Hence, we have

$$\dot{\mathbf{p}}^*(t) = -\mathbf{A}'\mathbf{p}^*(t), \qquad \mathbf{p}^*(T) = \mathbf{0} \quad \text{or} \quad \mathbf{p}^*(T) = \alpha^* \mathbf{v}^*, \tag{7.9}$$

$$\dot{\mathbf{x}}^*(t) = \mathbf{A}\mathbf{x}^*(t) + \mathbf{b}u^*(t), \qquad \mathbf{x}^*(0) = \mathbf{x}_0, \tag{7.10}$$

and

$$u^*(t) = \begin{cases} -\mathcal{E}_\rho\{\langle \mathbf{p}^*(t), \mathbf{b}\rangle\}, & p_0^* \neq 0, \\ -\operatorname{sgn}\{\langle \mathbf{p}^*(t), \mathbf{b}\rangle\}, & p_0^* = 0, \end{cases} \tag{7.11}$$

where

$$\mathcal{E}_\rho\{x\} \equiv \begin{cases} +1, & |x| \operatorname{sgn}\{x\} \geq 1, \\ -1, & |x| \operatorname{sgn}\{x\} \leq -1, \\ |x|(1/(\rho-1)) \operatorname{sgn}\{x\}, & |x| \leq 1. \end{cases} \tag{7.12}$$

Obviously, if $\rho > 1$, the problem is normal.

The case $p_0^* = 0$ can be dispensed with as in the minimum-fuel problem (see Lemma 6.2), and hence the initial-value problem which defines $\mathbf{M}(\mathbf{v}, \alpha)$ can be stated as follows.

Choose a $\mathbf{v} \in R_n$ and an $\alpha \in R_1^+$, let $\tau = T - t$, and solve the initial-value problem

$$\dot{\mathbf{x}}(\tau) = -\mathbf{A}\mathbf{x}(\tau) + \mathbf{b}\mathcal{E}_\rho\{\langle \alpha\mathbf{v}, e^{\mathbf{A}\tau}\mathbf{b}\rangle\}. \tag{7.13}$$

Definition 7.1

Let $\mathbf{M}$ be defined as

$$\mathbf{M}(\mathbf{v}, \alpha) = e^{-\mathbf{A}T}\{\mathbf{v} + \int_0^T e^{\mathbf{A}s}\mathbf{b}\,\mathcal{E}_\rho\{\langle \alpha\mathbf{v}, e^{\mathbf{A}s}\mathbf{b}\rangle\}\, ds\}. \tag{7.14}$$

7.3 Definitions and Preliminary Theorems

Once again the idea is to find the set of "effort-optimal" controls by finding the set D of elements $(\mathbf{v}, \alpha)$ which yield them. Since we intend to search the set of effort-optimal controls for the one that is feasible for the given state $\mathbf{x}_0$ we must know precisely which elements $(\mathbf{v}, \alpha)$ in R_{n+1} correspond to an effort-optimal control.

Lemma 7.1

If the terminal state $\mathbf{v}^*$ of a fixed-time energy-optimal trajectory is in the interior of S, then the corresponding optimal control is $u_{(0,T]}{}^* = 0$.

Proof:

The proof of Lemma 7.1 follows precisely that of Lemma 6.2.

As in Problem Z_3, it is also assumed here that $\|e^{\mathbf{A}T}\mathbf{x}_0\| > r$, and hence the solution is not the trivial one $u_{(0,T]}{}^* = 0$.

Definition 7.2 (D)

Let $D \subseteq \partial S \times R_1^+$ be the set of ordered pairs $(\mathbf{v}, \alpha)$, such that if $(\mathbf{v}, \alpha) \in D$, then $\mathbf{v} \in \partial S$ is the terminal state at $t = T$ of an optimal trajectory for Problem Z_4, that is, for some state in R_n/S.

Lemma 7.2

If $\|e^{\mathbf{A}T}\mathbf{x}_0\| > r$ (that is, if $u_{(0,T]} \neq 0$), if the matrix $-\mathbf{A}$ is positive definite, and if $T > T^*$, where T^* is the minimum time required to steer the state from $\mathbf{x}_0$ to ∂S, then the optimal control for Problem Z_4 is such that

$$J^* < T. \tag{7.15}$$

Proof:

The proof of Lemma 7.2 follows exactly the argument of the proof Lemma 6.1.

Lemma 7.2 means that the optimal control must be such that $|u(t)| < 1$ over some nonzero interval in $(0, T]$. Hence, the controls $u_{(0,T]} = 0$ and $u_{(0,T]} = \pm 1$ are ruled out. This is reasonable, since the former is the trivial solution and the latter uses more fuel than the minimum-time solution, and hence the engineer would not be interested in it. The mathematical reason for wishing to discard these control is that if an element $(\mathbf{v}, \alpha)$ gives rise to $u_{(0,T]} = +1, 0, -1$, the local effect of varying $\mathbf{v}$ and α will not alter $u_{(0,T]}$.

Lemma 7.3

1. If the plant of Problem Z_3 satisfies the normality conditions for Problem Z_1(Theorem C.1, Appendix C), then, for any $\mathbf{v}$ and $T > 0$, the control given by Equation 7.11 is not identically zero on $(0, T]$.
2. If the function $\langle \mathbf{v}, e^{\mathbf{A}\tau}\mathbf{b}\rangle$ has no zeros on the interval $(0, T)$, then there exists a number $\rho(\mathbf{v})$, such that if $\alpha \geqslant \rho(\mathbf{v})$, the ordered pair $(\mathbf{v}, \alpha) \notin D$.

Proof:

From Theorem C.1, we see that if the plant satisfies the normality conditions of Problem Z_1, then in any finite interval $(0, T)$ the function $\langle \mathbf{v}, e^{\mathbf{A}\tau}\mathbf{b}\rangle$ has at most a finite number of zeros which proves part 1, and

$$\rho(\mathbf{v}) = \frac{1}{\inf\limits_{\tau \in (0,T)} \{ |\mathbf{v}, e^{\mathbf{A}\tau}\mathbf{b}| \}} \tag{7.16}$$

as in Lemma 6.4 gives us part 2. Q.E.D.

Lemma 7.3 means that the best one can hope for is that $D = \partial S \times (0, \rho(\mathbf{v}))$. This situation is better than that encountered in Problem Z_3, where one has to contend with both a lower bound $\delta(\mathbf{v})$ and an upper bound $\rho(\mathbf{v})$ (see Lemma 6.4).

Theorem 7.1

If the matrix $-\mathbf{A}$ is positive definite, then $D = \partial S \times (0, \rho(\mathbf{v}))$.

Proof:

The proof of Theorem 7.1 can be obtained by replacing Equation 6.20 in Theorem 6.1 with

$$\| e^{\mathbf{A}T}\mathbf{M}(v, \alpha)\| = \| \mathbf{v} + \int_0^T e^{\mathbf{A}s}\mathbf{b}\, \mathcal{E}_\rho \{\langle \alpha \mathbf{v}, e^{\mathbf{A}s}\mathbf{b}\rangle\}\, ds \| , \tag{7.17}$$

and noting that $\delta(\mathbf{v}) = 0$ for Problem Z_4. Consider the function

$$J(\mathbf{v}, \alpha) = \int_0^T \frac{1}{\rho} \, | \mathcal{E}_\rho \{\langle \alpha \mathbf{v}, e^{\mathbf{A}s}\mathbf{b}\rangle\} |^\rho \, ds. \tag{7.18}$$

It can be shown, exactly as in Lemma 6.5, that $J(\mathbf{v}, \alpha)$ is continuous in $\mathbf{v}$ and α, and that given any fixed $\mathbf{v} \neq 0$, $J(\mathbf{v}, \alpha)$ is strictly increasing in α on $(0, \rho(\mathbf{v}))$.

Theorem 7.2

$\mathbf{M}(\mathbf{v}, \alpha)$ is one-to-one on D.

Proof:

It is shown that the extremal controls are unique in this problem in Appendix C. It follows that $\mathbf{M}(\mathbf{v}, \alpha)$ is one-to-one on D.

Consider the derivatives of $\mathcal{E}_\rho\{x\}$ (see Equation 7.12). Clearly, then,

$$\frac{d\mathcal{E}_\rho\{x\}}{dx} = \begin{cases} 0, & |x| > 1, \\ \frac{1}{\rho - 1} |x|^{(2-\rho)/(\rho-1)}, & |x| < 1. \end{cases} \tag{7.19}$$

In words, $\mathcal{E}_\rho\{x\}$ is piecewise differentiable, the derivative being discontinuous at $x = 1$. And we have

$$\frac{d|\mathcal{E}_\rho\{x\}|}{dx} = \begin{cases} 0, & |x| > 1 \\ \frac{1}{\rho - 1} |x|^{(2-\rho)/(\rho-1)} \operatorname{sgn}\{x\}, & |x| < 1. \end{cases} \tag{7.20}$$

Note that if $\rho > 2$, the derivatives of Equations 7.19 and 7.20 "blow up" for $x = 0$, and hence we must conclude that in order to differentiate $\mathbf{M}(\mathbf{v}, \alpha)$ and $J(\mathbf{v}, \alpha)$, we must restrict ρ to the interval $(1, 2]$. This fact was apparently unnoticed in Reference 31.

Theorem 7.3

If $\rho \in (1, 2]$, the function $\mathbf{M}(\mathbf{v}, \alpha)$ and $J(\mathbf{v}, \alpha)$ have continuous first partial derivatives to all order everywhere in D. Further, let $\eta(\mathbf{v}, \alpha)$, $\mu(\mathbf{v}, \alpha)$, $\tau_j(\mathbf{v}, \alpha)$, $j = 1, 2, \ldots, \mu(\mathbf{v}, \alpha)$. D_j, Q_μ, Q_η, Q_T, and Q be defined exactly as in Chapter 6. Then $\mathbf{M}(\mathbf{v}, \alpha)$ and $J(\mathbf{v}, \alpha)$ have continuous partial derivatives to all orders on every open set $D_i \subset D$.

Proof:

To obtain the first partial derivatives, we use Equations 7.14, 7.18, 7.19, and 7.20 as well as Definitions 6.3, 6.5, 6.6, and 6.11. Thus, we have

$$\begin{aligned} \frac{\partial J(\mathbf{v}, \alpha)}{\partial \alpha} &= \int_0^T \frac{1}{\rho} \frac{\partial}{\partial \alpha} |\mathcal{E}_\rho\{\langle \alpha \mathbf{v}, e^{\mathbf{A}s}\mathbf{b}\rangle\}|^\rho \, ds \\ &= \sum_{l=0}^{\xi} \int_{\tau_{2l+2-\eta}}^{\tau_{2l+3-\eta}} \frac{1}{\alpha(\rho - 1)} |\langle \alpha \mathbf{v}, e^{\mathbf{A}s}\mathbf{b}\rangle|^{\rho/(\rho-1)} \, ds, \end{aligned} \tag{7.21}$$

where, as in Theorem 6. 3,

$$\tau_0 \equiv 0, \tag{7.22}$$

$$\tau_{\mu+1} \equiv T, \tag{7.23}$$

$$2\xi + 3 - \eta = \mu + \frac{(1 + (-1)^{\mu+\eta})}{2}, \tag{7.24}$$

and $\tau_j(\mathbf{v}, \alpha)$, $\eta(\mathbf{v}, \alpha)$, and $\mu(\mathbf{v}, \alpha)$ have been written without their arguments for convenience. Similarly, we have

$$\frac{\partial J(\mathbf{v}, \alpha)}{\partial v_i} = \sum_{l=0}^{\xi} \int_{\tau_{2l+2-\eta}}^{\tau_{2+3-\eta}} \frac{1}{(\rho - 1)} |\langle \alpha\mathbf{v}, e^{\mathbf{A}s}\mathbf{b}\rangle|^{1/(\rho-1)}$$

$$\times \operatorname{sgn}\{\langle \alpha\mathbf{v}, e^{\mathbf{A}s}\mathbf{b}\rangle\} \langle \alpha\mathbf{i}, e^{\mathbf{A}s}\mathbf{b}\rangle \, ds, \tag{7.25}$$

where $\mathbf{i}$ is an n-vector whose elements are zero except for the ith which is 1. Also,

$$\frac{\partial \mathbf{M}(\mathbf{v}, \alpha)}{\partial \alpha} = e^{-\mathbf{A}T} \int_0^T e^{\mathbf{A}s}\mathbf{b} \frac{\partial}{\partial \alpha} \mathcal{E}_\rho \{\langle \alpha\mathbf{v}, e^{\mathbf{A}s}\mathbf{b}\rangle\} \, ds \tag{7.26}$$

$$= \sum_{l=0}^{\xi} e^{-\mathbf{A}T} \int_{\tau_{2l+2-\eta}}^{\tau_{2l+3-\eta}} e^{\mathbf{A}s}\mathbf{b}\, \alpha \frac{1}{(\rho - 1)}$$

$$\times |\langle \alpha, \mathbf{v}, e^{\mathbf{A}s}\mathbf{b}\rangle|^{1/(\rho-1)} \operatorname{sgn}\{\langle \alpha\mathbf{v}, e^{\mathbf{A}s}\mathbf{b}\rangle\} \, ds, \tag{7.27}$$

and

$$\frac{\partial \mathbf{M}(\mathbf{v}, \alpha)}{\partial v_i} = e^{-\mathbf{A}T} \left\{ \mathbf{i} + \sum_{l=0}^{\xi} \int_{\tau_{2l+2-\eta}}^{\tau_{2l+3-\eta}} e^{\mathbf{A}s}\mathbf{b} \frac{1}{(\rho - 1)} \right.$$

$$\times |\langle \alpha\mathbf{v}, e^{\mathbf{A}s}\mathbf{b}\rangle|^{(2-\rho)/(\rho-1)} \langle \alpha\mathbf{i}, e^{\mathbf{A}s}\mathbf{b}\rangle \, ds. \tag{7.28}$$

These sums, Equations 7. 21, 7. 25, 7. 27, and 7. 29, are obviously continuous on D_i. They are also continuous on Q, because, as $\eta(\mathbf{v}, \alpha)$ and $\mu(\mathbf{v}, \alpha)$ change (as $(\mathbf{v}, \alpha)$ is varied), the limits on the integrals in the sums vary so as to make the derivatives continuous. Now, the higher-order partial derivatives are not continuous on D, because, in taking these higher-order derivatives, one must find $\partial\tau_j(\mathbf{v}, \alpha)/\partial\alpha$ and $\partial\tau_j(\mathbf{v},\alpha)/\partial v_i$ (the limits on the integrals are functions of $\mathbf{v}$ and α).

These derivatives are given by Equations 6.52 and 6.53, and they do not exist on the set Q_μ. Similarly, since $\eta(\mathbf{v}, \alpha)$ and $\mu(\mathbf{v}, \alpha)$ are discontinuous in the sets Q_η and Q_μ, respectively, one concludes, in view of the fact that the integrands in Equations 7.21, 7.25, 7.27, and 7.28 have continuous partial derivatives for ρ in $(1, 2]$, that the theorem is demonstrated.

It can be seen that $\mathbf{M}(\mathbf{v}, \alpha)$ is "smoother" in Problem Z_4 if $\rho \in (1, 2]$, in that it is once differentiable everywhere. This is seemingly a consequence of the fact that the control law is continuous. The fact that one must restrict ρ to $(1, 2]$ is of small engineering significance.

Theorem 7.4

The $n \times n$ matrix $\mathbf{G}(\mathbf{v}, \alpha)$, defined in Definition 7.12, exists everywhere on D and has rank n.

Proof:

From Equation 7.25, we have

$$\mathrm{J}_{\mathbf{v}}(\mathbf{v},\alpha) = \sum_{l=0}^{\xi} \int_{\tau_{2l+2-\eta}}^{\tau_{2l+3-\eta}} \frac{\alpha}{(\rho-1)} |\langle \alpha\mathbf{v}, e^{\mathbf{A}s}\mathbf{b}\rangle|^{1/(\rho-1)}$$

$$\times \operatorname{sgn}\{\langle \alpha\mathbf{v}, e^{\mathbf{A}s}\mathbf{b}\rangle\} e^{\mathbf{A}s}\mathbf{b}\, ds. \tag{7.29}$$

From Equation 7.28, we obtain

$$\mathbf{M}_{\mathbf{v}}(\mathbf{v}, \alpha) = e^{-\mathbf{A}T} \left\{ \mathbf{I} + \sum_{l=0}^{\xi} \int_{\tau_{2l+2-\eta}}^{\tau_{2l+3-\eta}} \frac{1}{(\rho-1)} |\langle \alpha\mathbf{v}, e^{\mathbf{A}s}\mathbf{b}\rangle|^{(2-\rho)/(\rho-1)} \right.$$

$$\left. \times\, e^{\mathbf{A}s}\mathbf{b}\rangle\langle e^{\mathbf{A}s}\mathbf{b}\, ds \right\}. \tag{7.30}$$

Proceeding exactly as in the proof of Theorem 6.4, we have

$$\mathbf{G}(\mathbf{v}, \alpha) = \left\{ r\, e^{-\mathbf{A}T} [\mathbf{I} + \mathbf{E}] \mathbf{h}_\theta \,\Big|\, \frac{1}{\mathrm{J}_\alpha} \mathbf{M}_\alpha \right\}, \tag{7.31}$$

where, using Equations 6.61, 7.21, 7.26, 7.29, and 7.30, and letting

$$g(s) = \langle \alpha\mathbf{v}, e^{\mathbf{A}s}\mathbf{b}\rangle, \tag{7.32}$$

we have

$$f(s) = \langle \mathbf{x}, e^{\mathbf{A}s}\mathbf{b}\rangle, \tag{7.33}$$

$$\langle \mathbf{x}, \mathbf{E}\,\mathbf{x}\rangle = \frac{\alpha}{(\rho - 1)}\left\{ \sum_{l=0}^{\xi} \int_{\tau_{2l+2-\eta}}^{\tau_{2l+3-\eta}} |g(s)|^{(2-\rho)/(\rho-1)} f^2(s)\, ds \right.$$

$$- \frac{1}{\displaystyle\sum_{l=0}^{\xi} \int_{\tau_{2l+2-\eta}}^{\tau_{2l+3-\eta}} |g(s)|^{\rho/(\rho-1)}\, ds}$$

$$\left.\left(\sum_{l=0}^{\xi} \int_{\tau_{2l+2-\eta}}^{\tau_{2l+3-\eta}} f(s)\ |g(s)|^{1/(\rho-1)} \operatorname{sgn}\{g(s)\}\, ds\right)^2 \right\} \quad (7.34)$$

Since $g(s) = f(s)$ if $\mathbf{x} = \alpha\mathbf{v}$, we have

$$\langle \alpha\mathbf{v}, \mathbf{E}\,\alpha\mathbf{v}\rangle = 0, \quad (7.35)$$

and hence $\mathbf{E}$ cannot be definite. Clearly, $\mathbf{E}$ is positive semidefinite if and only if

$$\left(\sum_{l=0}^{\xi} \int_{\tau_{2l+2-\eta}}^{\tau_{2l+3-\eta}} |g(s)|^{(2-\rho)/(\rho-1)} |f(s)|^2 ds\right)$$

$$\times\left(\sum_{l=0}^{\xi} \int_{\tau_{2l+2-\eta}}^{\tau_{2l+3-\eta}} |g(s)|^{\rho/(\rho-1)}\, ds\right)$$

$$\geqslant\left(\sum_{l=0}^{\xi} \int_{\tau_{2l+2-\eta}}^{\tau_{2l+3-\eta}} f(s)\,|g(s)|^{1/(\rho-1)} \operatorname{sgn}\{g(s)\}\, ds\right)^2. \quad (7.36)$$

We let

$$\hat{g}(s) = \begin{cases} |g(s)|^{(\rho/2)/(\rho-1)}, & s \in [\tau_{2l+2-\eta}, \tau_{2l+3-\eta}], \\ \quad l = 0, 1, \ldots, \xi, & \\ 0, \text{otherwise}, & \end{cases} \quad (7.37)$$

$$\hat{f}(s) = \begin{cases} f(s)\,|g(s)|^{(1-\rho/2)/(\rho-1)} \operatorname{sgn}\{g(s)\}, & s \in [\tau_{2l+2-\eta}, \tau_{2l+3-\eta}], \\ \quad l = 0, 1, \ldots, \xi, & \\ 0, \text{otherwise}. & \end{cases} \quad (7.38)$$

Then Equation 7.36 becomes

$$\left(\int_0^T |\hat{f}(s)|^2 ds\right)\left(\int_0^T |\hat{g}(s)|^2 ds\right) \geq \left(\int_0^T |\hat{f}(s)\hat{g}(s)\right)^2, \tag{7.39}$$

which is Schwartz's Inequality (see [1]). The remainder of the proof is identical to that of Theorem 6.5. Since

$$\alpha \mathbf{v}' \mathbf{E}\, \mathbf{h}_\theta(\theta) = 0, \qquad \theta = \frac{1}{r}\, \mathbf{h}^{-1}(\mathbf{v}), \tag{7.40}$$

$$\alpha \mathbf{v}' \mathbf{E} = 0, \tag{7.41}$$

then, using Equations 7.21 and 7.26, we have

$$\frac{\langle \alpha \mathbf{v}, e^{\mathbf{A}T} \mathbf{M}_\alpha(\mathbf{v}, \alpha)\rangle}{J_\alpha(\mathbf{v}, \alpha)} = 1. \tag{7.42}$$

Q.E.D.

Note, as in Problems Z_1, Z_2, and Z_3, that the matrix **I** in Equation 7.31 (the effect of modifying the regulator problem so that the target is a hypersphere) is needed in order for the proof of Theorem 7.4 to succeed. Hence, in the modified problem, $\mathbf{G}(v, \alpha)$ has rank n everywhere, whereas in the unmodified problem it may not. This is the principle computational advantage of modifying the problems.

The multi-input case was not considered here. However, the method may be extended to include (1) m controllers, (2) $\mathbf{A} = \mathbf{A}(t)$, $\mathbf{b} = \mathbf{b}(t)$, and (3) $\mathbf{I} \to \mathbf{Q}$ in defining S.

7.4 The Iterative Procedure

The iterative procedure and initial-guess procedures for Problem Z_4 are described in Sections 6.4 and 6.6, respectively. In fact, Sections 6.4, 6.5, and 6.6 are also applicable to Problem Z_3. It can be concluded, in view of the fact that $\mathbf{G}(\mathbf{v}, \alpha)$ exists even on the set Q in this case, that the iterative procedure should be more successful in the neighborhood of this set than it was in both the time-optimal and the fuel-optimal case. Further, since $\mathbf{G}(\mathbf{v}, \alpha)$ is continuous, a theorem like Theorem 4.6 will apply (see [7]) when the guess is close enough, even if the point $(\mathbf{v}^*, \alpha)$ is in Q.

Despite the fact that $\mathbf{M}(v, \alpha)$ is "smoother" in Problem Z_4 than it is in Problems Z_1 and Z_3, there is little doubt that the iterations for Problem Z_4 will be more time consuming on a digital computer. This is because the evaluation of both $\mathbf{G}(\mathbf{v}, \alpha)$ and $\mathbf{M}(\mathbf{v}, \alpha)$ will involve the numerical evaluation of several complicated integrals (see Equations 7.21, 7.27, 7.29, and 7.30). No experiments were carried out to test the iterative procedure for Problem Z_4.

8. Experimental Results

8.1 Introduction

The iterative procedures of Sections 2.3, 3.4, 4.4, and 6.4 have been tested by the author both on the IBM 7094 digital computer at the Massachusetts Institute of Technology's Computation Center and on the disc version of the IBM 1620-II at the Computing Center of the Royal Military College of Canada. The latest version of these programs is given in Appendices F, G, H, and I. Several of the numerical experiments are described.

8.2 Experimental Results for Problem Z_5^1 (Section 2.3)

The program listed and described in Appendix H was used to carry out the experiments reported in this section. Three experiments are described here, and for each of these the plant **A** matrix was

$$\mathbf{A} = \begin{bmatrix} -.5 & 5. & 0. & 0. \\ -5. & -.5 & 0. & 0. \\ 0. & 0. & -.6 & 10. \\ 0. & 0. & -10. & -.6 \end{bmatrix},$$

which is the real Jordan canonical state-space representation for a plant having poles $(-0.5 \pm j5.)$ and $(-0.6 \pm j10.)$. In all cases, the **Q** matrix is assumed to be the 4×4 identity matrix. In the first two experiments, the target state was $\mathbf{z} = \mathbf{0}$ (that is, the origin), the initial state was $\mathbf{x}_0' = (10., 10., 10., 10)$, and $\mathbf{b}' = (0., 1., 0., 1.)$ (that is, one input) was used. Two inputs were used for the third experiment with $\mathbf{b}_1' = (0., 1., 0., 1.)$ and $\mathbf{b}_2' = (1., 0., 1., 0)$. The roles of the initial and target states were reversed [that is, $\mathbf{z}' = (10.,10.,10.,10)$, and $\mathbf{x}_0' = (0.,0.,0.,0.)$]. The sequence $\langle \mathbf{y}_k, \mathbf{Q}\mathbf{y}_k \rangle$ for the first two runs are plotted in Figure 8.1. In both of these tests, it happened that the fixed control interval was inadvertently chosen larger than the minimum time

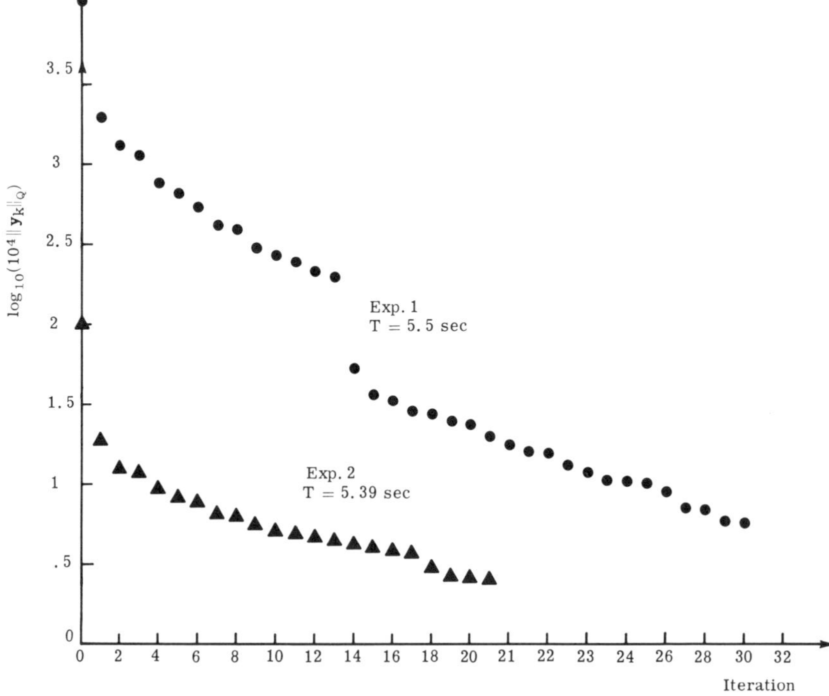

Figure 8.1 Experimental results for Problem $Z\frac{1}{5}$.

(T = 5.5 in Experiment 1 and T = 5.39 in Experiment 2). In both cases, the error was monotone decreasing (see Figure 8.1) and, as was pointed out in Section 2.3, converges toward a feasible control. It is apparent from Figure 8.1 that the rate of convergence is approximately exponential in that the semilog plots are approximately piecewise linear. Also, as will be shown in the sequel, the rate of convergence does not compare with that of the procedure described in Section 4.4, particularly when the guess is close in the case of the latter. Actually, one would expect the procedure of Section 4.4 to converge faster, since it uses the first two terms of the Taylor-series expansion of the error term $\|\mathbf{x}_0 - \mathbf{M}(\mathbf{v}_k, T_k)\|$.

In the third experiment, the control interval T = 5.5 was somewhat less than T*. This experiment was successful since the minimum was reached, to within the specified accuracy, in three iterations. These

Table 8.1—Problem Z_5^1 Example

Iteration number	0	1	2	3
$(\langle \mathbf{y}k, \mathbf{Q}\mathbf{y}_k\rangle)^{1/2}$	20.	17.9665	17.9642	17.9641
$(\langle(\bar{\mathbf{x}}_k(T) - \mathbf{z}, \mathbf{Q}(\bar{\mathbf{x}}_k(T) - \mathbf{z})\rangle)^{1/2}$	17.9665	17.9642	17.9641	17.9641

results are presented in Table 8.1. In this latter experiment, the number of switchings of the elements of $\bar{\mathbf{u}}_{(t_0,T]}$ did not change from iteration to iteration, whereas in the first two experiments the number of switchings changed significantly from iteration to iteration (that is, from 17 to 9 after one iteration).

8.3 Experimental Results for Problem Z_5^2 (Section 3.4)

Problem Z_5^2 is defined by choosing $\xi = \int_{t_0}^{T} \sum_{i=1}^{m} |u(s)|\, ds$ and $\Omega = (\mathbf{u}\colon |u_i| \leq 1, i = 1, 2, \ldots, m)$. The program in Appendix I was used to carry out the numerical experiments reported here.

It should perhaps be pointed out here that the process involved in finding $\bar{\mathbf{u}}_{k(t_0,T]}$ in Problem Z_5^2 is considerably more time consuming than that in Problem Z_5^1, since in the former case it is necessary to search for the α_k such that $\xi(\bar{\mathbf{u}}_{k(t_0,T]}) = \hat{\xi}$.

The plant, initial state, and target state used here were the same as those used in the first experiments of Section 8.3. Two runs were made to check if the solution in which the program of Appendix I was used was the same as that in which the program of Appendix H was used when the appropriate values were chosen for T and $\hat{\xi}$. The solutions for both programs coincided with one another. Some typical results are reported in Table 8.2 for $T = 4.3$ and $\hat{\xi} = 4.0$. Again, the

Table 8.2—Problem Z_5^2 Example

Iteration number	0	1	2	3	4
$(\langle \mathbf{y}_k, \mathbf{Q}\mathbf{y}_k\rangle)^{1/2}$	1.9652	0.9486	0.9439	0.9377	0.9342
$(\langle\bar{\mathbf{x}}_k(T) - \mathbf{z}),$ $\mathbf{Q}(\bar{\mathbf{x}}_k(T) - \mathbf{z})\rangle)^{1/2}$	0.9486	1.0438	0.9379	0.2975	0.9319
Number of switchings	14	26	22	14	19

error sequence is monotone decreasing, and the structure of the extremal controls (that is, the number of switchings) varies considerably from iteration to iteration.

8.4 Experimental Results for Problem Z_1 (Section 4.4)

The program used to solve Problem $Z_1(\Omega_a)$ is listed in Appendix F.

The minimization indicated in Equation 4.79 is only approximately carried out in the program. The following three steps describe the procedure used in the approximation:

Step 1. Choose $\gamma_k = 1$, and solve the initial-value problem (that is, evaluate $\mathbf{M}(\mathbf{v}_{k+1}(\gamma_k = 1), T_{k+1}(\gamma_k = 1))$, and find

$$\hat{\mathbf{e}}_{k+1} = \mathbf{e}_{k+1}(\gamma_k = 1). \tag{8.1}$$

Step 2. If $\|\hat{\mathbf{e}}_{k+1}\| < \frac{1}{2}\|\mathbf{e}_k\|$, index k and proceed to the next iteration. If, on the other hand, $\|\hat{\mathbf{e}}_{k+1}\| \geq \frac{1}{2}\|\mathbf{e}_k\|$, solve the initial-value problem using

$$\tilde{\gamma}_k = \frac{1}{2}\,\frac{\|\mathbf{e}_k\|}{\|\hat{\mathbf{e}}_{k+1}\|} \tag{8.2}$$

to find $\tilde{\mathbf{e}}_{k+1}$.

Step 3. If $\|\tilde{\mathbf{e}}_{k+1}\| < \|\mathbf{e}_k\|$, proceed to the next iteration. If, on the other hand, $\|\tilde{\mathbf{e}}_{k+1}\| \geq \|\mathbf{e}_k\|$, solve the initial-value problem using

$$\gamma_k = \frac{\tilde{\gamma}_k{}^2\|\mathbf{e}_k\|}{2(\|\tilde{\mathbf{e}}_{k+1}\| - (1 - \tilde{\gamma}_k)\|\mathbf{e}_k\|)}. \tag{8.3}$$

This procedure is motivated by Relation 4.92, and it assumes that

$$\|\mathbf{e}_{k+1}(\gamma_k)\| \cong (1 - \gamma_k)\,\|\mathbf{e}_k\| + \gamma_k{}^2 N_k. \tag{8.4}$$

The idea is to evaluate N_k by choosing a γ_k. Step 1 assumes that $N_k = \|\hat{\mathbf{e}}_{k+1}\|$. The choice of $\frac{1}{2}$ is motivated by the fact that the minimum of the right-hand side of Equation 8.3 occurs at

$$\gamma_k = \frac{1}{2}\,\frac{\|\mathbf{e}_k\|}{N_k}. \tag{8.5}$$

If it turns out that $\|\hat{\mathbf{e}}_{k+1}\| \nless \|\mathbf{e}_k\|$, then in Step 3 we assume that N_k satisfies the equation

$$\|\tilde{\mathbf{e}}_{k+1}\| = |1 - \tilde{\gamma}_k|\,\|\mathbf{e}_k\| + \tilde{\gamma}_k{}^2 N_k;$$

hence, the choice of Equation 8.3. This procedure could be carried still further with γ_k getting still smaller if the error were not reduced. However, one reaches the point where the changes being made in the guess are smaller than the machine accuracy. The arbitrary choice of three steps worked well in application.

A series of some thirty experiments were carried out with various plants of order 4(n = 4) having both real and complex eigenvalues. (An "experiment" means the solution of one problem, that is, a given plant and a given initial state.) In each case, the purpose of the experiment was either to test the behavior of the iterative procedure near the set Q_μ or to test the initial-guess procedure of Section 4.6, or both. For complex eigenvalues, the separated problem was a second-order plant with complex conjugate eigenvalues. In this case, rather than actually solving the second-order time-optimal problem, we used the minimum-time solution for the following problem as an estimate:

$$\dot{y}(t) = 2\lambda y(t) + c, \tag{8.6}$$

$$y(0) = \|\mathbf{x}_0\|^2, \tag{8.7}$$

$$y(T) = r^2, \tag{8.8}$$

where λ is the real part of the complex eigenvalue, and

$$c = \|\mathbf{b}_i\|^2, \tag{8.9}$$

and where $\mathbf{b}_i$ is the control vector for the separated second-order plant. That is, we have

$$T_{est} = \frac{1}{|2\lambda|} \log \left\{ \frac{\frac{c}{|2\lambda|} + \|\mathbf{x}_0\|^2}{\frac{c}{|2\lambda|} + r^2} \right\}. \tag{8.10}$$

This procedure approximates the minimum isochrones of the second-order plant by circles.

Four experiments are discussed in some detail, and the behavior of the sequences $\|\mathbf{e}_k\|$ and $T^* - T_k$ are displayed for these experiments in Figures 8.2 through 8.5, along with an illustration of how the controls varied during the iteration.

Experiment 1 (Figure 8.2)

Purpose: To test the initial-guess procedure and the approximate scheme for choosing γ_k.

Plant:

$$\mathbf{A} = \begin{bmatrix} -0.5 & 5.0 & 0 & 0 \\ -5.0 & -0.5 & 0 & 0 \\ 0 & 0 & -0.6 & 10.0 \\ 0 & 0 & -10.0 & -0.6 \end{bmatrix}, \quad \mathbf{b} = \begin{bmatrix} 0 \\ 1 \\ 0 \\ 1 \end{bmatrix}, \quad r = 1. \tag{8.11}$$

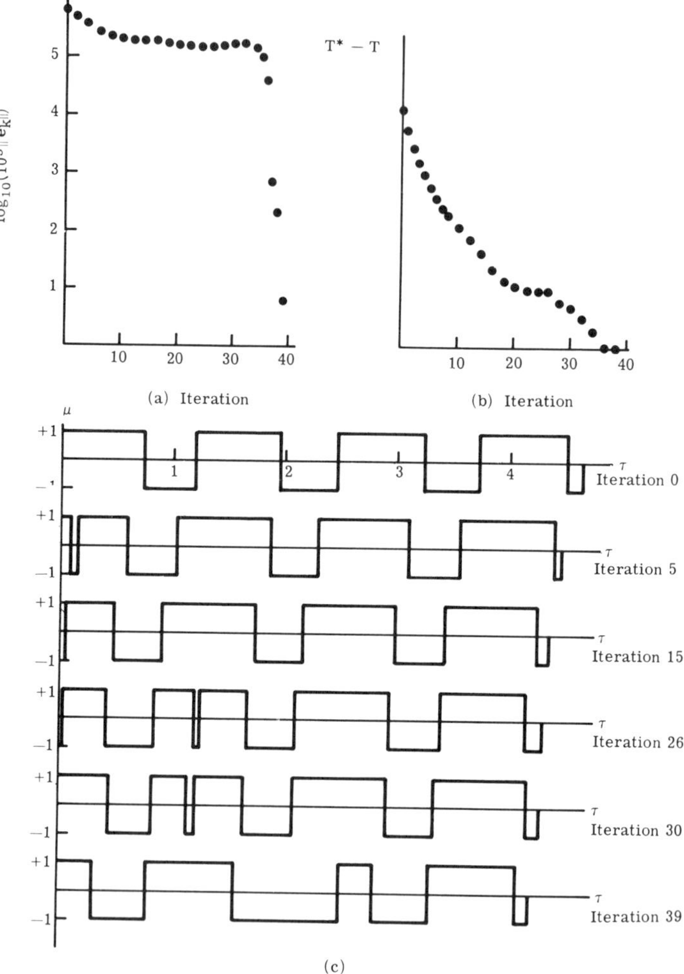

Figure 8.2 Experiment 1 (τ is reverse time).

Data:

$\mathbf{x}_0$	$\mathbf{v}_0$	$\mathbf{M}(\mathbf{v}_0, T_0)$	$\mathbf{v}_{39}$	$\mathbf{M}(\mathbf{v}_{39}, T_{39})$
10.0	−0.814	11.5	0.096	10.0
10.0	0.229	15.2	−0.747	10.0
10.0	0.195	10.4	−0.618	10.0
10.0	−0.518	7.0	0.223	10.0

(8.12)

$T_0 = 4.61, \qquad T_{39} = 4.402$

Remarks:

1. The computation time was 41 sec (7094 time).
2. This experiment illustrates how the sequence $(\mathbf{v}_k, T_k)$ can traverse the set Q (that is, go from one set D_i to another) with an almost monotone decreasing error. (Note that the error plot is $\log_{10}\{10^5\|\mathbf{e}_k\|\}$ vs. iteration number.) The change from one set D_i to another is characterized by a change in the number of switchings which went from 7 to 9 to 8 to 10 to 9 to 7 in this example (see Figure 8.2).
3. The initial-value problem was solved 83 times, and the matrix $\mathbf{G}$ and its inverse $\mathbf{G}^{-1}$ were calculated 37 times.
4. $\|\mathbf{e}_k\|$ decreased through five orders of magnitude in 39 iterations.

Experiment 2 (Figure 8.3)

Purpose: To test the initial-guess procedure when the time estimate is large.[1]

Plant:

$$\mathbf{A} = \begin{bmatrix} -1. & 0 & 0 & 0 \\ 0 & -2. & 0 & 0 \\ 0 & 0 & -3. & 0 \\ 0 & 0 & 0 & -4. \end{bmatrix}, \quad \mathbf{b} = \begin{bmatrix} 1. \\ 2. \\ 3. \\ 4. \end{bmatrix}, \quad r = 1. \qquad (8.13)$$

Data:

$\mathbf{x}_0$	$\mathbf{v}_0$	$\mathbf{M}(\mathbf{v}_0, T_0)$	$\mathbf{v}_{13}$	$\mathbf{M}(\mathbf{v}_{13}, T_{13})$
3.	0.993	6.97	0.873	3.0
−1.	−0.827	13.68	−0.484	−1.0001
4.	0.083	68.3	0.0394	3.9997
1.	0.005	256.3	−0.0499	0.99916

(8.14)

$T_0 = 1.3863, \quad T_{13} = 0.9191$

Remarks:

1. The computation time was approximately 18 sec (7094 time).
2. The structure of the control varied considerably (see Figure 8.2), again showing movement of the sequence $(\mathbf{v}_k, T_k)$ through several sets D_i.
3. The initial-value problem was solved 28 times, and the matrices $\mathbf{G}(\mathbf{v}, \alpha)$ and $\mathbf{G}^{-1}(\mathbf{v}, \alpha)$ were evaluated 11 times.

---

[1] We used $T_{est} = \max_i ((1/|\lambda_i|) \log(1. + |x_{0i}|))$ rather than $\max_i ((1/|\lambda_i|) \log(1 + |x_{0i}|)/(1 + r)))$.

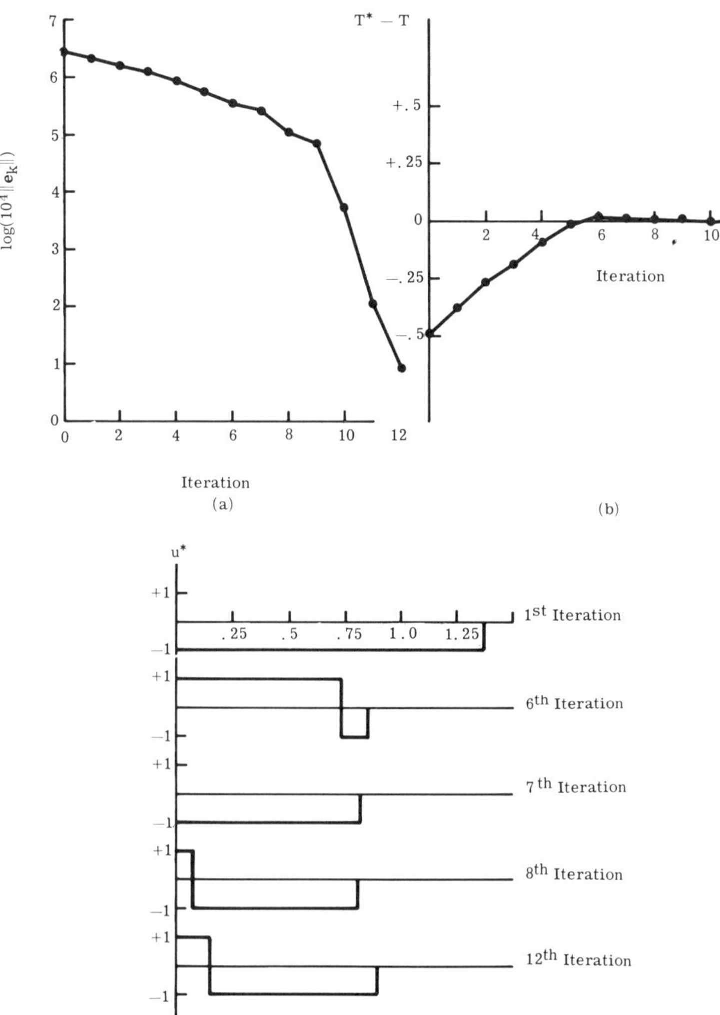

Figure 8.3 Experiment 2.

4. The error $\|e_k\|$ decreased through six orders of magnitude in 13 iterations.

Experiment 3 (Figure 8.4)

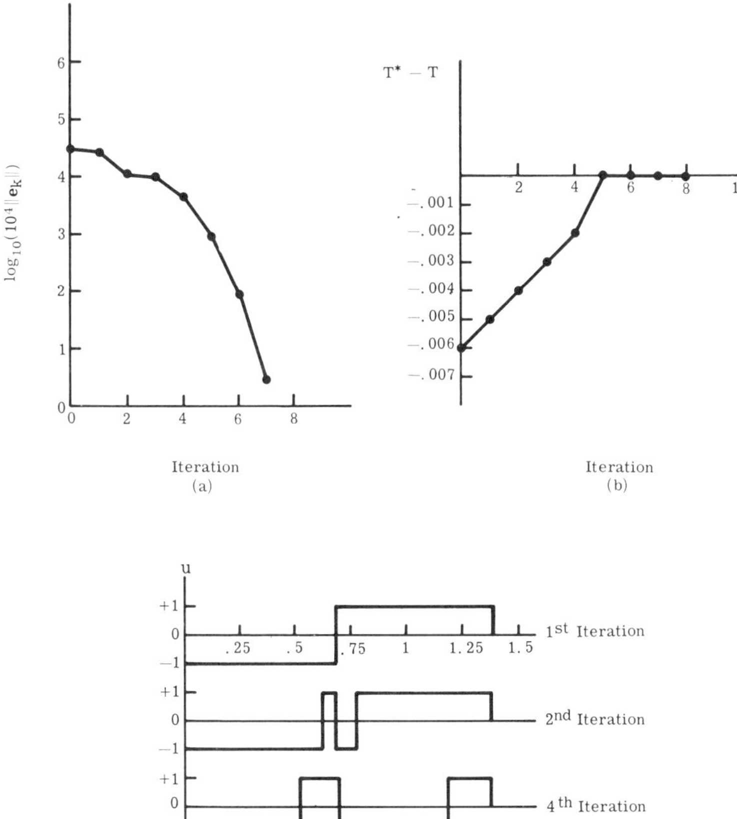

Figure 8.4 Experiment 3.

Purpose: To investigate the behavior of the iterative procedure near a point in Q_μ (that is, near the set upon which the derivatives "blow up").

Method: Two arbitrary times, $\hat{T} = 1.386$ and $\hat{\tau}_j = 0.693$, were chosen. Then (see the explanation of SUBROUTINE WORST in Appendix F), $\hat{\mathbf{v}}$ was calculated in such a way that

$$\langle \hat{\mathbf{v}}, \mathbf{A}^l e^{\mathbf{A}\hat{\tau}_j}\mathbf{b} \rangle = 0, \quad l = 0, 1, 2. \tag{8.15}$$

The state $\hat{\mathbf{x}} = \mathbf{M}(\hat{\mathbf{v}}, \hat{T})$ was calculated, and then $\mathbf{x}_0$ was chosen by adding the scalar 1. to each element of $\hat{\mathbf{x}}$. The initial guess was then chosen as $(\hat{\mathbf{v}}, \hat{T})$. This procedure was used in order to find an initial guess as close to the set Q_μ as single-precision arithmetic will allow.

Plant:

$$\mathbf{A} = \begin{bmatrix} -1. & 0 & 0 & 0 \\ 0 & -2. & 0 & 0 \\ 0 & 0 & -3. & 0 \\ 0 & 0 & 0 & -4. \end{bmatrix}, \quad \mathbf{b} = \begin{bmatrix} 1. \\ 2. \\ 3. \\ 4. \end{bmatrix}, \quad r = 1. \tag{8.16}$$

Data:

$\hat{\mathbf{v}}$	$\mathbf{M}(\hat{\mathbf{v}}, \hat{T})$	$\mathbf{x}_0$	$\mathbf{v}_7$	$\mathbf{M}(\mathbf{v}_7, T_7)$
−0.18262	0.2938	1.2938	−0.14367	1.2938
0.54778	17.851	18.851	0.50320	18.851
−0.73027	2.548	3.548	−0.74961	3.548
0.36508	318.85	319.85	0.4058	319.85

(8.17)

$\hat{T} = 1.396, \quad \hat{\tau}_j = 0.693, \quad T_7 = 1.3801$

Remarks:

1. As can be seen, the iterative procedure behaved very well near the set Q_μ. We say "near" because the subroutine calculating the switch times (see the explanation of SUBROUTINE TIMES 5 in Appendix F) does not calculate the switch times exactly. Because of this, instead of having $\langle \mathbf{v}, \mathbf{A}e^{\mathbf{A}\tau_j}\mathbf{b}\rangle = 0$, it was of the order of 10^{-8}.
2. Figure 8.4 demonstrates how a switch time satisfying Equations 8.15 evolves into three switchings very close together when the vector $\mathbf{v}$ is changed slightly.
3. The initial-value problem was solved 13 times, and the matrices $\mathbf{G}$ and $\mathbf{G}^{-1}$ were evaluated 4 times.
4. The computation time was approximately 10 sec (7094 time).

Experiment 4 (Figure 8.5)

Purpose: To investigate the behavior when the point $(\mathbf{v}^*, T^*) \in Q_\mu$. Also the plant was chosen so as to be difficult to control (that is, poles close together).

Method: Two arbitrary times, $\hat{T} = 3$, and $\hat{\tau}_j = 1.5$, were chosen, and the point $\hat{\mathbf{v}}$ was chosen as in Experiment 3. Then, $\mathbf{x}_0 = \mathbf{M}(\hat{\mathbf{v}}, \hat{T})$ was calculated, and the initial-guess procedure of Section 4.6 was used to start the iterative procedure.

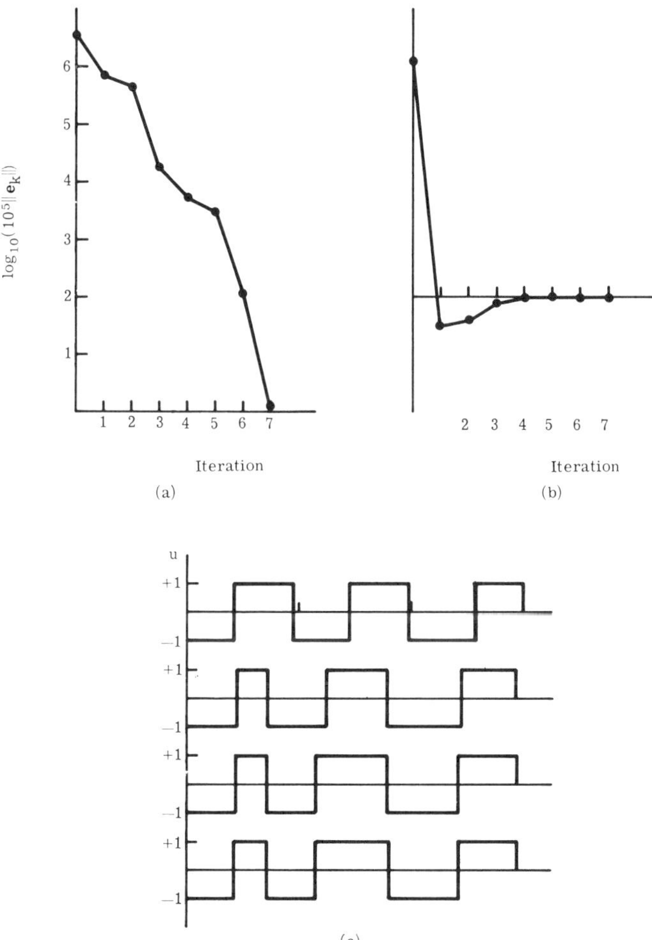

Figure 8. 5 Experiment 4 (τ is reverse time).

Plant:

$$\mathbf{A} = \begin{bmatrix} -1.0 & 4.9 & 0 & 0 \\ -4.9 & -1.0 & 0 & 0 \\ 0 & 0 & 1.5 & 4.76 \\ 0 & 0 & -4.76 & -1.5 \end{bmatrix}, \quad \mathbf{b} = \begin{bmatrix} 0 \\ 1 \\ 0 \\ 1 \end{bmatrix}, \quad r = 1. \tag{8.18}$$

Data:

$\mathbf{M}(\hat{\mathbf{v}}, \hat{T})$	$\mathbf{v}_0$	$\mathbf{M}(\mathbf{v}_0, T_0)$	$\mathbf{v}_7$	$\mathbf{M}(\mathbf{v}_7, T_7)$
5.7891	−0.5176	−3.378	−0.5199	5.7891
−9.3492	0.0043	−13.49	−0.2496	−9.3492
−46.937	0.7623	−77.81	0.6883	−46.937
66.579	0.3884	65.57	0.4400	66.579

(8.19)

$\hat{T} = 3, \quad \hat{\tau}_j = 1.5, \quad T_0 = 3, \quad T_7 = 2.958$

Remarks:

1. Note that in this experiment the initial-time guess T_0 was very close to the solution T_7. It is probably true that the closeness of this guess is the reason for the total number of iterations required here being somewhat less than that required for Experiment 1, where the T_0 was not close.
2. The initial-value problem was solved 15 times, and the matrices $\mathbf{G}$ and $\mathbf{G}^{-1}$ were evaluated 2 times.
3. The computation time was approximately 6 sec (7094 time).

The following data are presented without figures.

Experiment 5

Plant:

$$\mathbf{A} = \begin{bmatrix} -1.0 & 4.9 & 0 & 0 \\ -4.9 & -1.0 & 0 & 0 \\ 0 & 0 & -3. & 4. \\ 0 & 0 & -4. & -3. \end{bmatrix}, \quad \mathbf{b} = \begin{bmatrix} 0 \\ 1 \\ 0 \\ 1 \end{bmatrix}, \quad r = 1. \tag{8.20}$$

Purpose: The same as in Experiment 4.

Data:

$\mathbf{M}(\hat{\mathbf{v}}, \hat{T})$	$\mathbf{v}_0$	$\mathbf{M}(\mathbf{v}_0, T_0)$	$\mathbf{v}_9$	$(\mathbf{v}_9, T_9)$
−12.175	0.206	−21.8	−0.0365	−12.175
−0.986	0.3932	0.594	−0.0490	−0.986
8271.0	0.361	7708.5	0.3759	8270.9
5534.5	0.820	5114.1	0.9246	5534.5

$T_0 = 3.00, \quad T_9 = 3.00, \quad \|\mathbf{e}_0\| = 702.2 \quad \|\mathbf{e}_9\| = 0.00012$

Number of evaluations of $\mathbf{M} = 17$

Number of evaluations of $\mathbf{G}$ and $\mathbf{G}^{-1} = 4$,

Computation time ≈7 sec (7094).

Experiment 6

Plant: The same plant as in Experiment 5.

Purpose: To test procedure with $r = 10$.

Data:

$\mathbf{x}_0$	$\mathbf{v}_0$	$\mathbf{M}(\mathbf{v}_0, T_0)$	$\mathbf{v}_8$	$\mathbf{M}(\mathbf{v}_8, T_8)$
100.	0.307	0.436	0.299	100.
200.	−9.99	199.91	−9.54	200.
300.	0.032	−112.7	0.004	300.
−400.	0.032	−2674.9	0.153	−400.

(8.21)

$\|\mathbf{e}_0\| = 2312$, $\|\mathbf{e}_8\| = 0.00007$, $T_0 = 3.105$, $T_8 = 3.050$

Number of evaluations of $\mathbf{M} = 16$,

Number of evaluations of $\mathbf{G}$ and $\mathbf{G}^{-1} = 4$,

Computation time ≈ 6.3 sec (7094).

Experiment 7

Plant:

$$A = \begin{bmatrix} -1.0 & 4.9 & 0 & 0 \\ -4.9 & -1.0 & 0 & 0 \\ 0 & 0 & -4.0 & 3.0 \\ 0 & 0 & -3.0 & -4.0 \end{bmatrix}, \quad \mathbf{b} = \begin{bmatrix} 0 \\ 1 \\ 0 \\ 1 \end{bmatrix}, \quad r = 1 \tag{8.22}$$

Data:

$\mathbf{x}_0$	$\mathbf{v}_0$	$\mathbf{M}(\mathbf{v}_0, T_0)$	$\mathbf{v}_{360}$	$\mathbf{M}(\mathbf{v}_{360}, T_{360})$
10.	0.716	18.35	0.697	10.
20.	−0.621	19.39	0.690	20.
30.	−0.289	7980.8	−0.066	30.
−40.	0.137	−22293.0	−0.186	−40.

(8.23)

$\|\mathbf{e}_0\| = 2.36 \times 10^4$, $T_0 = 2.877$, $\|\mathbf{e}_{360}\| = 5.7 \times 10^{-5}$, $T_{360} = 2.631$

Remarks:

This was the longest run of all the experiments carried out. The computation time was about 4.75 (7094) minutes. The large number of iterations is attributed to the poor guess. The error sequence was monotone decreasing throughout.

Experiment 8

Plant:

$$\mathbf{A} = \begin{bmatrix} -1.0 & 4.9 & 0 & 0 \\ -4.9 & -1.0 & 0 & 0 \\ 0 & 0 & -2.0 & 4.58 \\ 0 & 0 & -4.58 & -2.0 \end{bmatrix}, \quad \mathrm{b} = \begin{bmatrix} 0 \\ 1 \\ 0 \\ 1 \end{bmatrix}, \quad \mathrm{r} = 1. \tag{8.24}$$

Purpose: Same as for Experiment 1.

Data:

$\mathbf{x}_0$	v_0	$\mathbf{M}(\mathrm{v}_0, \mathrm{T}_0)$	v_{10}	$\mathbf{M}(\mathrm{v}_{10}, \mathrm{T}_{10})$
10.	0.840	13.18	0.7533	10.
20.	−0.528	25.02	0.5356	20.
30.	−0.116	110.3	−0.0571	30.
40.	−0.14	−11.2	−0.3773	−40.

(8.25)

$\|\mathbf{e}_0\| = 85.49, \quad \|\mathbf{e}_{10}\| = 0.00009, \quad \mathrm{T}_0 = 2.905, \quad \mathrm{T}_{10} = 2.668$

Number of evaluations of $\mathbf{M} = 14$,

Number of evaluations of $\mathbf{G}$ and $\mathbf{G}^{-1} = 3$,

Computation time ≈6 sec (7094).

Experiment 9

Plant: The same plant as in Experiment 8.

Purpose: See Experiment 3.

Data: $\hat{\mathrm{T}} = 3$

$\mathbf{M}(\hat{\mathrm{v}}, \hat{\mathrm{T}})$	v_0	$\mathbf{M}(\mathrm{v}_0, \mathrm{T}_0)$	v_{10}	$\mathbf{M}(\mathrm{v}_{10}, \mathrm{T}_{10})$
−1.596	−0.195	−4.12	−0.183	−1.596
4.487	−0.06	−0.168	−0.094	4.487
−94.00	0.824	0.819	0.823	−94.00
520.7	0.528	0.518	0.529	520.7

(8.26)

$\|\mathbf{e}_0\| = 2.0, \quad \|\mathbf{e}_{10}\| = 0.00005, \quad \mathrm{T}_0 = 3. , \quad \mathrm{T}_{10} = 3.001$

Computation time ≈5 sec (7094).

Remarks:

The plants of Experiments 4 through 9 were chosen so that the norm of the steady-state state vector in response to a unit step $\|\mathbf{A}\mathbf{b}^{-1}\|$ was constant. This was achieved by using the same **b** vector for each plant and choosing the eigenvalues to lie on a circle of given radius (5 in this case) in the s-plane. One pair of poles was kept fixed for all the plants, and the other poles were selected so as to have a class of plants in which the poles were varying distances apart. The reason for doing this is that the controllability matrix becomes more nearly singular as the poles come closer together. Hence, testing such a class of plants might show the effects of "controllability" on the initial-guess procedure and/or the iterative procedure. Besides the experiments reported here (that is, 4 through 9), another set of experiments was also carried out where the initial state was $\mathbf{x}_0' = (10.,10.,10.,10)$ for each plant. The results showed that the initial-guess procedure and the iterative procedure worked equally well for all plants in this class.

Experiment 10

Plant:[2]

$$\mathbf{A} = \begin{bmatrix} -2. & 2. & 0 & 0 \\ -2. & -2. & 0 & 0 \\ 0 & 0 & -1. & 10. \\ 0 & 0 & -10. & -1. \end{bmatrix}, \quad \mathbf{b} = \begin{bmatrix} 0 \\ 2 \\ 0 \\ 1 \end{bmatrix}.$$

Purpose: To compare the iterative procedure with r = 1.0, 0.1, 0.01, and 0.001.

Data: r = 1.0

$\mathbf{x}_0$	$\mathbf{v}_0$	$\mathbf{M}(\mathbf{v}_0, T_0)$	$\mathbf{v}_{13}$	$\mathbf{M}(\mathbf{v}_{13}, T_{13})$
10.	−0.049	20.74	−0.159	10.
10.	−0.071	15.70	0.248	10.
10.	0.129	13.30	−0.836	10.
10.	0.988	12.54	−0.463	10.

$\|\mathbf{e}_0\| = 12.85$, $T_0 = 2.448$, $\|\mathbf{e}_{13}\| = 0.5 \times 10^{-5}$, $T_{13} = 2.224$

Number of evaluations of $\mathbf{M}$ = 27,

Number of evaluations of $\mathbf{G}$ and $\mathbf{G}^{-1}$ = 11,

Computation time ≈13 sec (7094).

r = 0.1

$\mathbf{x}_0$	$\mathbf{v}_0$	$\mathbf{M}(\mathbf{v}_0, T_0)$	$\mathbf{v}_{10}$	$\mathbf{M}(\mathbf{v}_{10}, T_{10})$
10.	0.00231	−19.16	0.0158	10.
10.	0.00448	−21.69	0.00333	10.
10.	−0.0688	9.847	−0.0404	10.
10.	0.0724	9.274	0.0900	10.

---

[2] This plant was also studied in Reference 22.

$\|\mathbf{e}_0\| = 43.07$, $T_0 = 2.987$, $\|\mathbf{e}_{10}\| = 5 \times 10^{-6}$, $T_{10} = 3.026$,
Number of evaluations of $\mathbf{M} = 10$,
Number of evaluations of $\mathbf{G}$ and $\mathbf{G}^{-1} = 2$,
Computation time $\approx$5 sec (7094).

$r = 0.01$

$\mathbf{x}_0$	$\mathbf{v}_0$	$\mathbf{M}(\mathbf{v}_0, T_0)$	$\mathbf{v}_{40}$	$\mathbf{M}(\mathbf{v}_{40}, T_{40})$
10.0	2.4×10^{-4}	-18.34	2.68×10^{-3}	10.0
10.0	4.4×10^{-4}	-15.75	-3.38×10^{-3}	10.0
10.0	-6.1×10^{-3}	8.78	9.00×10^{-3}	10.0
10.0	7.9×10^{-3}	8.03	5.15×10^{-4}	10.0

$\|\mathbf{e}_0\| = 38.4$, $T_0 = 2.997$, $\|\mathbf{e}_{40}\| = 6.7 \times 10^{-5}$, $T = 3.217$
Number of evaluations of $\mathbf{M} = 121$,
Number of evaluations of $\mathbf{G}$ and $\mathbf{G}^{-1} = 37$,
Computation time $\approx$49 sec (7094).

$r = 0.001$

$\mathbf{x}_0$	$\mathbf{v}_0$	$\mathbf{M}(\mathbf{v}_0, T_0)$	$\mathbf{v}_{39}$	$\mathbf{M}(\mathbf{v}_{39}, \mathbf{T}_{39})$
10.0	2.4×10^{-5}	-18.44	2.50×10^{-4}	10.0
10.0	4.4×10^{-4}	-15.80	-3.28×10^{-4}	10.0
10.0	-6.1×10^{-3}	8.65	9.09×10^{-4}	10.0
10.0	7.9×10^{-3}	7.91	-5.56×10^{-5}	10.0

$\|\mathbf{e}_0\| = 38.5$, $T_0 = 2.997$, $\|\mathbf{e}_{39}\| = 2.83 \times 10^{-5}$, $T_{39} = 3.23$,
Number of evaluations of $\mathbf{M} = 119$,
Number of evaluations of $\mathbf{G}$ and $\mathbf{G}^{-1} = 36$,
Computation time $\approx$48 sec (7094).

Remarks:

1. From the data for $r = 0.01$ and $r = 0.001$ in Experiment 10, it can be seen that the optimal terminal states for these two cases have almost the same directions and that the optimal times are the same to within less than $\frac{1}{3}\%$. As the target set radius is reduced still further, the differences between the successive solutions become negligibly small. We can conclude that the targets of radius 0.01 or 0.001 would yield "engineering" time-optimal solutions to the problem of hitting the origin.
2. The fact that the last two problems in Experiment 10 (that is, $r = 0.01$ and 0.001) required considerably more computation time than the first two is probably because the initial

guess was poor and not because the radius of the target hypersphere was small. From the data, as shown by $M(\mathbf{v}_0, T_0)$, it is clear that $-\mathbf{v}_0$ would have been almost as good a guess as $\mathbf{v}_0$ itself. In both of these computations, as in Experiment 7 where $r = 1.0$, the error sequence decreased rapidly for the first six or seven iterations and then decreased rather slowly for a time as the set Q_η was approached. This set was then crossed with an increase in the error (an arbitrary lower bound of 5×10^{-4} was chosen for γ_k–see subroutine CALNG in Appendix F), after which the error converged rapidly toward zero (that is, almost an order of magnitude decrease per iteration). The set Q_η is recognizable from the switch times because, when the set is crossed, the control $u_{k(0,T]}$ gains or loses a switch time very close to $\tau = 0$. Because the plant is stable, the integration of the initial-value problem in reverse time is very sensitive to changing the switch times near $\tau = 0$. Hence, small changes in the guesses are necessary near the set Q_η.

3. The sensitivity of the iterative procedure to the size of the target hypersphere will probably depend not only on the particular plant but also on the initial state $\mathbf{x}_0$. As we have seen, the effect of the hyperspherical target is to remove the corners from the isochrones. Then, when the current guess is such that $\mathbf{M}(\mathbf{v}_k, T_k)$ is near what would have been a corner in the original problem, the iterative procedure will undoubtedly behave better for r large. However, in regions where the isochrones are essentially smooth and flat, the behavior is unaffected by r.

4. Experiment 7 and the last two problems of Experiment 10 illustrate that the procedure can work quite well even when the initial guess is rather bad. Experiment 7 is particularly interesting. In this run, an initially rapid error reduction was achieved by essentially reducing the time to about one third of the minimum time. A long period then followed, during which the time was monotonically increased and the error sequence remained monotonically decreasing, but very slowly, until the proper set D_i was reached and the guess was "close." The error sequence then went rapidly toward zero. The number of switchings changed from 5 to 3 to 2 to 3 to 4 during the computations.

8.5 Experimental Results for Problem Z_3 (Section 6.4)

The program of Appendix F was adopted for use in the fixed-time–fuel-optimal problem, and this modified program is included in Appendix G.

The program was written to test both the iterative and the initial-guess procedures. Approximately thirty fuel-optimal problems were solved for four different plants, two of which were second-order, and two fourth-order. For three of the plants, a series of solutions for the same initial state but for different control intervals were carried out to determine the sensitivity of the minimum fuel to the control interval. These results are reported, as well as results similar to those found for the time-optimal case.

In all cases, the initial guess for the terminal point on the hypersphere was chosen as

$$\mathbf{v}_1 = \frac{r e^{\mathbf{A}T}\mathbf{x}_0}{\|e^{\mathbf{A}T}\mathbf{x}_0\|}; \tag{8.27}$$

this appears to have been a favorable guess. In most of the cases, the choice of α such that

$$\langle \mathbf{v}_1, e^{\mathbf{A}T}(\mathbf{x}_0 - \mathbf{M}(\mathbf{v}_1, \alpha_1)) \rangle = 0 \tag{8.28}$$

was very favorable, in that a solution was achieved in four iterations on the average. Also, it was found that 16 evaluations of **M** were required on the average to find the α_1 approximately satisfying Equation 8.28. It was found that for the given initial state, for each plant studied, there was a small range of times for which the zero of Equation 8.28 was difficult to find because the scalar product $\langle \mathbf{v}, e^{\mathbf{A}T}(\mathbf{x}_0 - \mathbf{M}(\mathbf{v}, \alpha)) \rangle$ was very sensitive to changes in α near the zero. Since the program allows only a fixed number of choices for α, the zero was not found accurately, with the result that the initial guess for J was not close. In these cases, the error sequence was monotone decreasing throughout, but after 30 iterations was not sufficiently small. This occurred for the time of 4.7 in the plant of Figure 8.7 and for the time of 3.2 in the plant of Figure 8.6, and the error behavior was similar to that which occurred in Experiment 7.

In general, the computation time per iteration for the minimum-fuel program is approximately the same (only slightly longer) as that required for the minimum-time program.

Experiment 1 (Figure 8.6)

Plant:

$$\mathbf{A} = \begin{bmatrix} -2. & 2. \\ -2. & -2. \end{bmatrix}, \quad \mathbf{b} = \begin{bmatrix} 0 \\ 2. \end{bmatrix}, \quad r = 0.1. \tag{8.29}$$

Purpose:

To find the fuel-cost-versus-time curve for the initial state $\mathbf{x}_0'(10,10)$.

Method:

The minimum-time problem was solved first, and then T was chosen in increments so as to obtain the points on the graph shown in Figure 8.6.

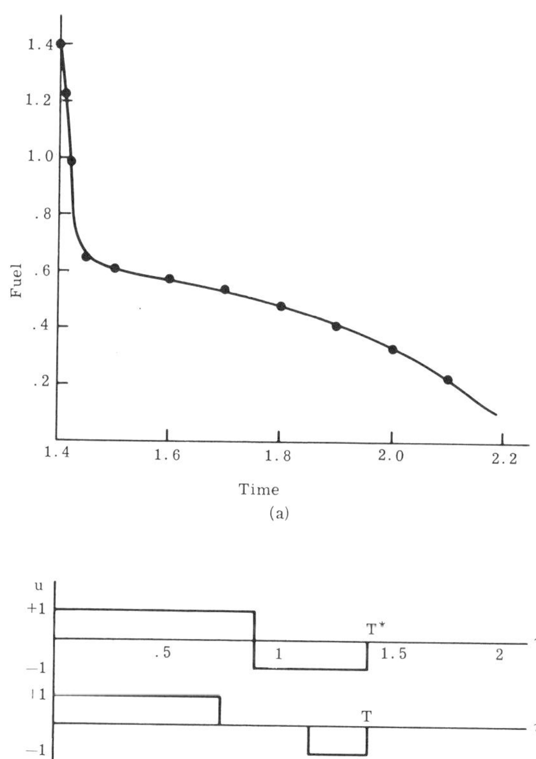

Figure 8.6 Effect of T on J* and $u^*_{(0,T]}$ (Equation 8.29).

Remarks:

The curve J vs. T and some solutions $u^*_{(0,T]}$ are given in Figures 8.6(a) and (b), respectively. It is noted that for small increases in the control interval, the fuel cost can be greatly reduced up to a point, after which lower fuel costs are achieved only at the expense of relatively large increases in the control interval. Results similar to this were also reported in Reference 11 for the suboptimal design of a fourth-order plant having two inputs.

Experiment 2 (Figure 8.7)

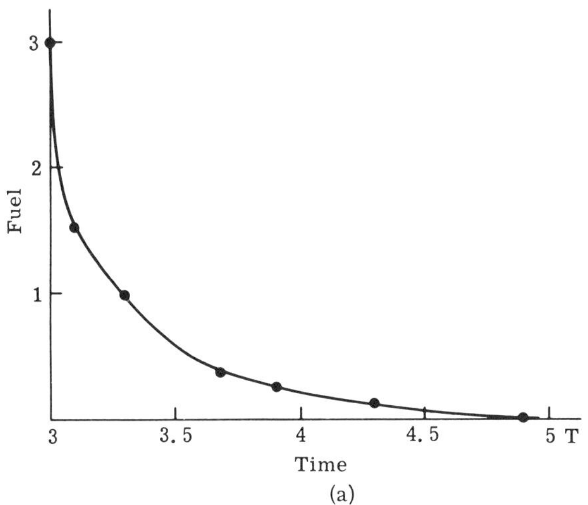

(a)

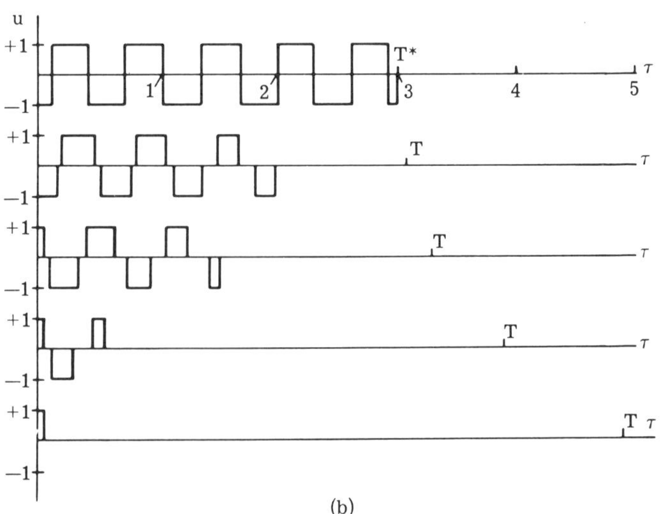

(b)

Figure 8.7 Effect of T on J^* and $u^*_{(0,T]}$ (Equation 8.30).

Plant:

$$\mathbf{A} = \begin{bmatrix} -1 & 10 \\ -10 & -1 \end{bmatrix}, \quad \mathbf{b} = \begin{bmatrix} 0 \\ 1 \end{bmatrix}, \quad r = 0.1. \tag{8.30}$$

Purpose: See Experiment 1.

Remarks:

This experiment had the same purpose and essentially the same results as Experiment 1 (see Figure 8.7).

Experiment 3 (Figure 8.8)

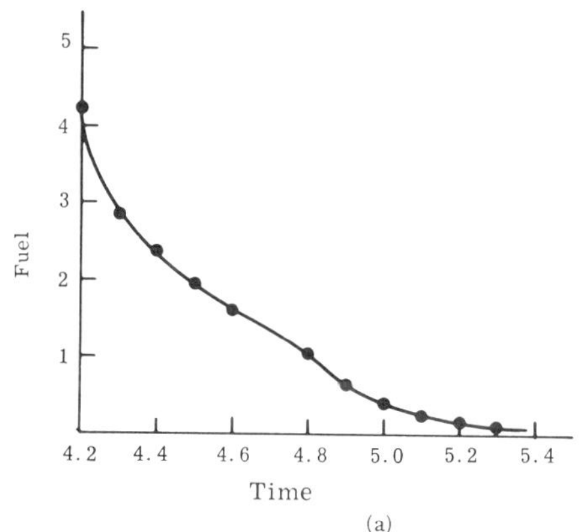

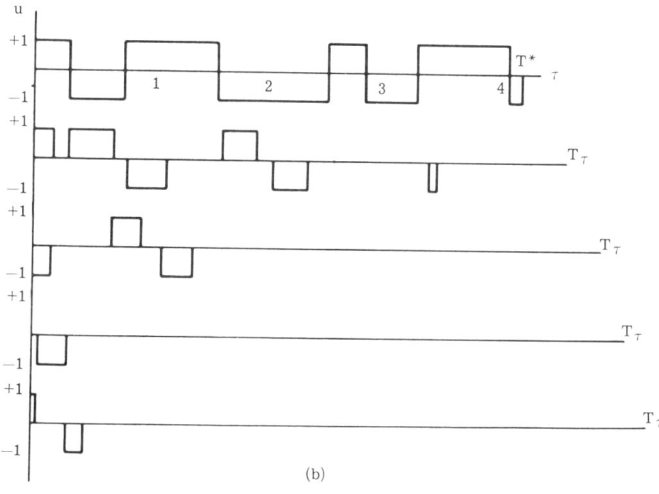

Figure 8.8 Effect of T on J* and $u^*_{(0,T]}$ (Equation 8.31).

Plant:

$$\mathbf{A} = \begin{bmatrix} -0.5 & 5 & 0 & 0 \\ -5 & -0.5 & 0 & 0 \\ 0 & 0 & -0.6 & 10 \\ 0 & 0 & -10 & -0.6 \end{bmatrix}, \quad \mathbf{b} = \begin{bmatrix} 0 \\ 0.5 \\ 0 \\ 0.6 \end{bmatrix}, \quad r = 1.0. \qquad (8.31)$$

Purpose: See Experiment 1

Remarks: This fourth-order problem has a similar $J \sim T$ curve.

The fact that each of the three experiments resulted in the same type of $J \sim T$ curve can best be seen by examining Figure 8.9, where all

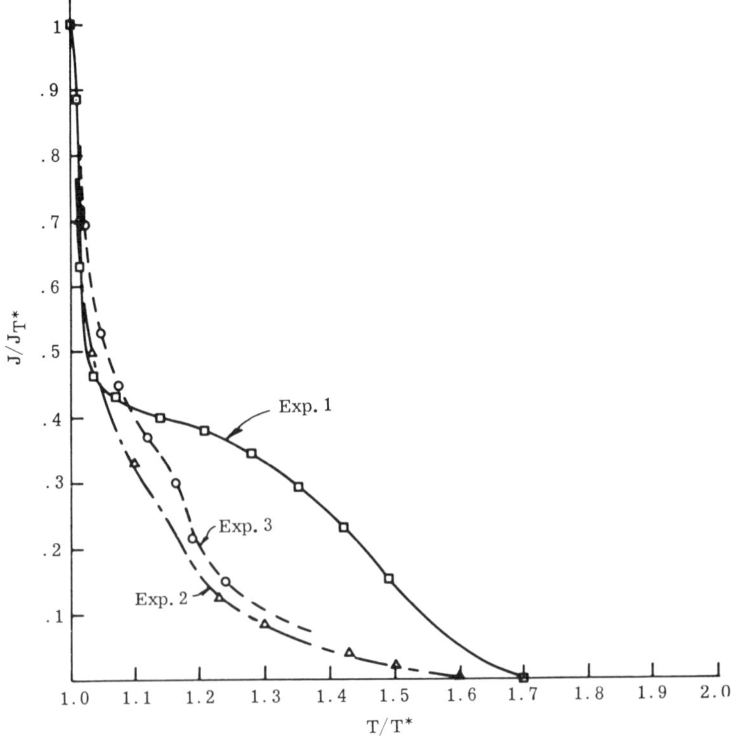

Figure 8.9 Trade-off between time and fuel.

three curves have been normalized and J/J_{T*} has been plotted versus T/T^*, where T^* is the minimum time and J_{T*} is the fuel used by the minimum-time control. These curves are quite similar, especially in their initial steep descent from the minimum-time point. This indicates that the minimum-time criterion is not a good "engineering" criterion. These curves may be worth generating for a given plant for a large number of initial conditions, in order to choose a desirable control interval as a fixed multiple of the minimum time.

The error sequence, the $J^* - J_k$ and $\alpha^* - \alpha_k$ sequences, and some of the controls are given in Figures 8.10a, b, and c, respectively for the plant of Experiment 3 and a control interval of T = 4.3. The first initial-guess procedure of Section 6.6 was used, and the computation time was 15 sec on the 7094. Except for the case mentioned previously in the preamble to this section when the control interval was chosen at 4.7, this particular computer run was one of the longer ones.

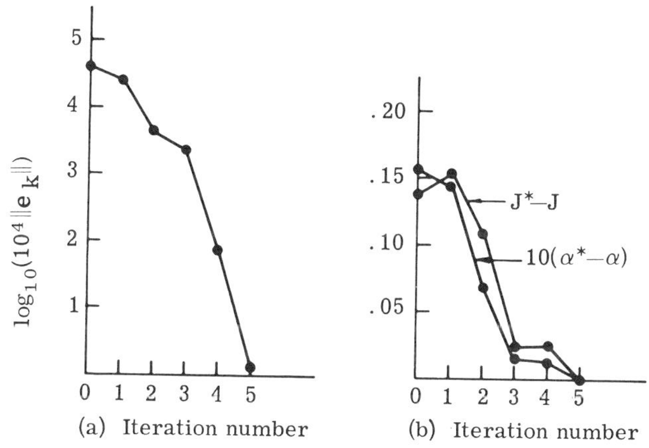

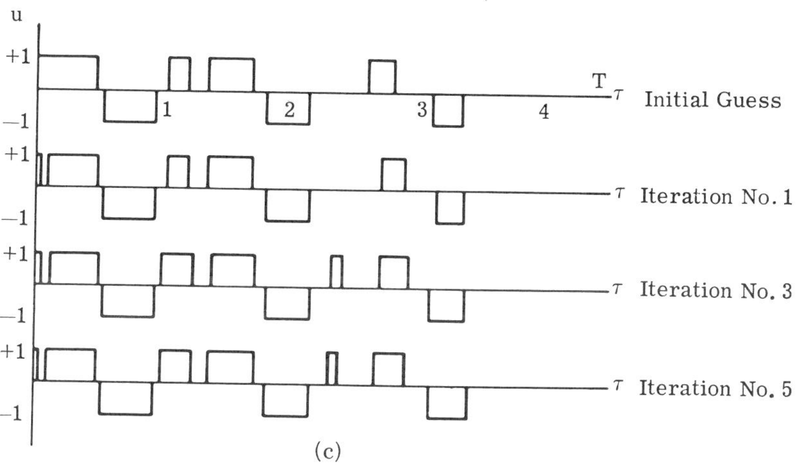

Figure 8.10 Typical error sequence.

Appendix A. Point Set-Theory Concepts

A definite collection of objects viewed as a single object is called a set. The objects are called elements and are said to be of the, in the, or to belong to the set. The set is said to contain the elements. The fact that x is an element of S is written $x \in S$. A set is well defined if, given any object, it is possible to determine if the object is an element of the set. The statement, "S is the set of all objects x having the property P" is usually written

$$S = \{x{:}x \text{ has property } P\}.$$

A new set, called a subset, can be formed from a given set by collecting all or some of the objects in the given set. If S_t is a subset of S, we write

$$S_t \subset S.$$

Two sets S_t and S are equal if and only if both $S_t \subset S$ and $S \subset S_t$, in which case we write

$$S_t = S.$$

A set containing no elements is said to be empty.

Definition A. 1. Ordered pair

An ordered pair (x_1, x_2) is a set such that $(x_1, x_2) = (y_1, y_2)$ if and only if $x_1 = y_1$ and $x_2 = y_2$.

Definition A. 2. Cartesian product

Given two sets S and R, the set of all ordered pairs (x, y), such that $x \in S$ and $y \in R$, is called the Cartesian product of S and R and is written $S \times R$.

Definition A. 3. Relation

A relation is a set of ordered pairs. If S is a set of ordered pairs and if $(x, y) \in S$, then "x is in the relation S to y."

Definition A. 4. Domain

The domain of a relation S, written $D_0\{S\}$, is the set of x having the property that there exists a y such that $(x, y) \in S$.

Definition A. 5. Range

The range of a relation S, written $R_a\{S\}$, is the set of y's for which there exists an x, such that $(x, y) \in S$.

Definition A. 6. Function

A function F is a relation having the special property that $(x, y) \in F$ and $(x, z) \in F$ imply that $y = z$. The statement y is a function of x is written

$$y = F(x). \tag{A. 1}$$

A function is defined by specifying a rule such as Equation A. 1 and a set of x's in $D_0\{F\}$.

Let F be a function with $D_0\{F\}$ and let $S \subset D_0\{F\}$. The set of F(x) such that $x \in S$ is called "the image of S under F" and is written as F[S]. Let $F[S] \subset T$, then F is said to be a mapping of S into T. If $F[S] = T$, then F is said to map S onto T. A mapping of a set S onto itself is called a transformation.

Definition A. 7. One-to-one

Let F be a function and let $S \subset D_0\{F\}$. The statement "F is one-to-one on S" means that for every x_1 and x_2 in S, $F(x_1) = F(x_2)$ implies $x_1 = x_2$.

Definition A. 8. Converse

Let S be a relation. The new relation $\check{S}$ defined

$$\check{S} = \{(x, y) : (y, x) \in S\}$$

is called the converse of S.

Definition A. 9. Inverse

Let a relation F be a function. If the converse of **F** is also a function, then $\check{F}$ is called the inverse of F and is written F^{-1}. It is said that F possesses an inverse or that the inverse of F exists.

Theorem A. 1

If a function is one-to-one on $S \subset D_0\{F\}$, then $\check{F}$ is a function on F[S].

Proof:

It is shown that $\check{F}$ satisfies Definition A. 6. Now, $(x, y) \in \check{F}$ means that $(y, x) \in F$. Similarly, $(x, z) \in \check{F}$ means that $(z, x) \in F$. Thus $F(y) = F(z)$, and, since F is one-to-one, $y = z$. Hence, $\check{F}$ is a function.

Definition A. 10. Union

Let S_1 and S_2 be two sets. The set of elements that are either in S_1 or in S_2 are in the union of S_1 and S_2, written $S_1 \cup S_2$.

Definition A. 11. Intersection

Let S_1 and S_2 be two sets. The set of elements that are in both S_1 and S_2 are in the intersection of S_1 and S_2, written $S_1 \cap S_2$.

Definition A. 12. Complement

The complement of S_1 relative to S_2 is the set of elements in S_2 but not in S_1 and is written S_2/S_1.

Definition A. 13. Distance

A distance $d(x, y)$, $x \in S$ and $y \in S$, is a function from $S \times S$ into the positive real line such that we have

1. $d(x, y) \geq 0$ for all x and all y in S. Also $d = 0$ if and only if $x = y$.
2. $d(x, y) = d(y, x)$ for all x and y in S.
3. $d(x, z) \leq d(x, y) + d(y, z)$ for all x, all y, and all z in S.

Definition A. 14. Continuous

Let F be a function on S and let $d(x, y)$ be a distance function on S. Then, F is continuous on S if the statement, "for every $\epsilon > 0$, there exists a $\delta > 0$ (depending only on ϵ) such that if $x \in S$ and $y \in S$ and $d(x, y) < \delta$, then $d(F(x), F(y)) < \epsilon$" holds.

Definition A. 15. Neighborhood

Let $x_0 \in S$. Then, the set of all points such that $d(x_0, x) < r$ is called an r open neighborhood of x and is denoted by $N(x_0, r)$.

Definition A. 16. Interval

Let a and b be two real numbers with $a \leq b$. The set of points on the real line between and perhaps including a and b is called an interval. Intervals can be closed, open, or half closed. Thus, we have the following:

1. $(a, b) = \{x : a < x < b\}$, open,
2. $(a, b] = \{x : a < x \leq b\}$, half closed,
3. $[a, b] = \{x : a \leq x \leq b\}$, closed.

Definition A. 17. Interior point

A point x in a set S is called an interior point of S if there exists an $r > 0$ such that $N(\mathbf{x}, r) \subset S$. The set S is said to be open if all of its points are interior points.

Definition A. 18. Accumulation point

A point x in S is an accumulation point of S if, for all $r > 0$, $N(x, r)$ contains at least one element of S distinct from x.

Definition A. 19. Euclidean n-space

The set of all real n vectors is called the Euclidean n-space R_n.

Definition A. 20. Bounded set in R_n

Let $S \subset R_n$. Then, S is bounded if there exists an $r > 0$ such that $S \subset N(0, r)$.

Definition A. 21. Closed in R_n

Let $S \subset R_n$. Then, S is closed if it contains all its accumulation points.

Definition A. 22. Compact in R_n

Let $S \subset R_n$. Then, S is compact if it is both closed and bounded.

Definition A. 23. Convex

Let $S \subset R_n$. Then, S is convex, if, given any two elements $\mathbf{x}$ and y in S and any α in $[0, 1]$, $\alpha\mathbf{x} + (1 - \alpha)\mathbf{y}$ is also in S.

Appendix B. n-Dimensional Geometry

Definition B. 1. Vector

A real n-vector is an ordered set of real numbers. Column vectors are used, that is

$$\mathbf{x} = \begin{bmatrix} x_1 \\ x_2 \\ \cdot \\ \cdot \\ \cdot \\ x_n \end{bmatrix} . \tag{B. 1}$$

The number n is called the dimension of x, and a row vector $\mathbf{x}'$ is called the transpose of **x**.

Definition B. 2. Linear vector space

A linear vector space V is a set of vectors such that we have the following:

1. if **x** and **y** are in V, then the vector $(\mathbf{x} + \mathbf{y})$ is also in V;
2. the vector sum is commutative and associative in V;
3. there is a unique vector **0** in V, such that $\mathbf{x} + \mathbf{0} = \mathbf{x}$ for all x in **V**;
4. for each **x** in V, there is also a vector $-\mathbf{x}$, such that $\mathbf{x} - \mathbf{x} = \mathbf{0}$;
5. to every scalar α and every **x** in **V**, there is a vector $\alpha\mathbf{x}$ in V.

Example B. 1. The Euclidean n-space is a linear vector space.

Definition B. 3. Scalar product

Let **x** and **y** be two vectors in R_n. Then, the scalar product $\langle \mathbf{x}, \mathbf{y} \rangle$ of **x** and $\mathbf{y}$ is defined as

$$\langle \mathbf{x}, \mathbf{y} \rangle = \sum_{i=1}^{n} x_i y_i, \tag{B.2}$$

and

$$1. \quad \langle \mathbf{x}, \mathbf{y} \rangle = \langle \mathbf{y}, \mathbf{x} \rangle, \tag{B.3}$$

$$2. \quad \langle \alpha\mathbf{x}, \mathbf{y} \rangle = \alpha \langle \mathbf{x}, \mathbf{y} \rangle, \tag{B.4}$$

$$3. \quad \langle \mathbf{x} + \mathbf{y}, \mathbf{z} \rangle = \langle \mathbf{x}, \mathbf{z} \rangle + \langle \mathbf{y}, \mathbf{z} \rangle, \tag{B.5}$$

$$4. \quad |\langle \mathbf{x}, \mathbf{y} \rangle| \leqslant \langle \mathbf{x}, \mathbf{x} \rangle^{1/2} \langle \mathbf{y}, \mathbf{y} \rangle^{1/2}. \tag{B.6}$$

Definition B.4. Norm

Let $\mathbf{x} \in R_n$. Then, the norm of $\mathbf{x}$, denoted by $\|\mathbf{x}\|$, is a function from R_n onto the positive real line having the following properties:

$$1. \quad \|\mathbf{x}\| \geqslant 0 \quad \text{and} \quad = 0 \quad \text{iff} \quad \mathbf{x} = \mathbf{0}; \tag{B.7}$$

$$2. \quad \|\alpha\mathbf{x}\| = |\alpha| \, \|\mathbf{x}\|, \qquad \alpha \text{ a scalar}; \tag{B.8}$$

$$3. \quad \|\mathbf{x}_1 + \mathbf{x}_2\| \leqslant \|\mathbf{x}_1\| + \|\mathbf{x}_2\|. \tag{B.9}$$

There are many norms. For example, a family of norms can be defined as follows:

$$\|\mathbf{x}\|_\rho = \left(\sum_{i=1}^{m} |x_i|^\rho \right)^{1/\rho}, \qquad 1 \leqslant \rho \leqslant \infty. \tag{B.10}$$

As ρ goes to infinity, $\|\mathbf{x}\|_\rho$ approaches the supremum or sup norm which is defined as

$$\|\mathbf{x}\|_\infty = \sup\, (|x_i|)_{i=1}{}^{n}. \tag{B.11}$$

Definition B.5. Linear transformation

A linear transformation is a transformation which possesses the superposition property. That is, we let F be a linear transformation on S and let $\mathbf{x}$, $a\mathbf{x}$, $\mathbf{y}$, $b\mathbf{y} \in S$, where a and b are scalars. Then, we have

$$F(a\mathbf{x} + b\mathbf{y}) = a\,F(\mathbf{x}) + b\,F(\mathbf{y}). \tag{B.12}$$

Definition B.6. Matrix

Let T be a linear function from R_n into R_m: that is, $\mathbf{y} = T(\mathbf{x})$, $\mathbf{y} \in R_m$ and $\mathbf{y} \in R_n$, defined as

$$y_i = \sum_{j=1}^{n} a_{ij} x_j, \qquad i = 1, 2, \ldots, n. \tag{B.13}$$

Then, T has a representation called a matrix, denoted by a boldface capital letter, which is an m × n array of the real coefficients in Equation 1.22). Thus,

$$\mathbf{A} \equiv \begin{bmatrix} a_{11} & a_{12} & \cdots & a_{1n} \\ a_{21} & & & \cdot \\ \cdot & & & \\ \cdot & & & \cdot \\ \cdot & & & \\ & & & \cdot \\ a_{m1} & a_{m2} & \cdots & a_{mn} \end{bmatrix}. \tag{B.14}$$

The mapping is written

$$\mathbf{y} = \mathbf{A}\mathbf{x}, \quad \mathbf{x} \in R_n, \quad \mathbf{y} \in R_m, \tag{B.15}$$

where the meaning of Equation B.15 is given in Equation B.18.

Definition B.7. Linear independence

A set of n vectors $\mathbf{x}_i$, i = 1, 2, n, are said to be linearly independent if the sum

$$\sum_{i=1}^{n} \alpha_i \mathbf{x}_i = \mathbf{0}, \quad \alpha_i \in R_1 \tag{B.16}$$

implies that

$$\alpha_i = 0, \quad i = 1, 2, \ldots, n.$$

Definition B.8. Bases

Let $\mathbf{a}_j$, j = 1, 2, ..., n be n vectors and let $\mathbf{y} \in R_n$. Then, the set of vectors $\{\mathbf{a}_j\}_1^n$ is called a basis for R_n if every $\mathbf{y} \in R_n$ can be written as a linear combination of the $\mathbf{a}_j$; that is,

$$\mathbf{y} = \sum_{j=1}^{n} \mathbf{a}_j x_j, \tag{B.17}$$

or, what is equivalent, let the n × n matrix **A** have columns $\mathbf{a}_j$. The columns of **A** form a basis, and the matrix **A** is said to have rank n if the equation

$$\mathbf{A}\mathbf{x} = \mathbf{y} \tag{B.18}$$

has a solution for all **y**.

Definition B.9. Quadratic form

The quadratic form of a matrix **A** is a function from $R_n \times R_n$ into R_1, defined as

$$\langle \mathbf{x}, \mathbf{Ax} \rangle = \langle \mathbf{A}'\mathbf{x}, \mathbf{x} \rangle = \sum_{i=1}^{n} \sum_{j=1}^{n} a_{ij} x_i x_j. \tag{B.19}$$

Definition B. 10. Positive definite

A square matrix **A** is said to be positive definite if

$$\langle \mathbf{x}, \mathbf{Ax} \rangle \geq 0, \tag{B.20}$$

and

$$\langle \mathbf{x}, \mathbf{Ax} \rangle = 0 \quad \text{implies that} \quad \mathbf{x} = 0.$$

Furthermore, **A** is said to be positive semidefinite if

$$\langle \mathbf{x}, \mathbf{Ax} \rangle \geq 0, \tag{B.21}$$

but

$$\langle \mathbf{x}, \mathbf{Ax} \rangle = 0 \quad \text{does not imply that} \quad \mathbf{x} = 0.$$

Theorem B. 1

If **A** or $-\mathbf{A}$ is positive definite, then $\mathbf{A}^{-1}$, the inverse of A, exists.

Definition B. 11. Dyad (outer product)

Let **x** and **y** be two elements in R_n. Then, the outer product (or dyad) of **x** and **y**, denoted by $\mathbf{x}\rangle\langle\mathbf{y}$, is the n × n matrix

$$\mathbf{x}\rangle\langle\mathbf{y} \equiv \begin{bmatrix} x_1y_1 & x_1y_2 & \cdots & x_1y_n \\ x_2y_1 & & & \cdot \\ \cdot & & & \cdot \\ \cdot & & & \cdot \\ \cdot & & & \cdot \\ x_ny_1 & x_ny_2 & \cdots & x_ny_n \end{bmatrix}. \tag{B.22}$$

Definition B. 12. Vector-valued function

A function whose range is in R_n is said to be vector-valued and will be denoted by a boldface lower-case letter followed by a list of characters denoting the elements of the domain in brackets.

Definition B. 13. Derivative of a vector

Let $\mathbf{x}(t)$ be a vector-valued function of the single variable t. Then, the derivative of $\mathbf{x}(t)$ with respect to t, written $\dot{\mathbf{x}}(t)$ or $\mathbf{x}_t(t)$, is a vector whose coordinates are the derivatives of the coordinates of $\mathbf{x}(t)$.

Definition B. 14. Derivatives (Frechet)

Let E be an open set in R_n, let **f** be a function which maps E into R_m, and let $\mathbf{x} \in E$. Then, **f** is said to be differentiable in the sense of Frechet if there exists a linear operator, denoted here by $f_x(x)$, such that

$$\lim_{h \to o} \frac{\|\mathbf{f}(\mathbf{x} + \mathbf{h}) - \mathbf{f}(\mathbf{x}) - \mathbf{f_x}(\mathbf{x})\mathbf{h}\|}{\|\mathbf{h}\|} = 0, \tag{B. 23}$$

where $\mathbf{h} \in R_n$ and $\mathbf{x} + \mathbf{h} \in E$.

Theorem B. 2

If $(\partial \mathbf{f}(\mathbf{x})/\partial x_i)$, $i = 1, 2, \ldots, n$ exist and are continuous, then and only then is **f** differentiable in the sense of Frechet ([15] p. 192), and we have

$$f_x(x) = \begin{bmatrix} \frac{\partial f_1(\mathbf{x})}{\partial x_1} & \frac{\partial f_1(\mathbf{x})}{\partial x_2} & \cdots & \frac{\partial f_1(\mathbf{x})}{\partial x_n} \\ \cdot & & & \cdot \\ \cdot & & & \cdot \\ \cdot & & & \cdot \\ \frac{\partial f_m(\mathbf{x})}{\partial x_1} & \frac{\partial f_m(\mathbf{x})}{\partial x_2} & \cdots & \frac{\partial f_m(\mathbf{x})}{\partial x_n} \end{bmatrix}, \quad (m \times n). \tag{B. 24}$$

Definition B. 14 and Theorem B. 2 are extendible to higher derivatives. For example, we can regard $\mathbf{f_x}(\mathbf{x})$ as a function mapping an open set E in R_n into $R_m \times R_m$. The second derivative of **f** at **x** is denoted $\mathbf{f_{xx}}(\mathbf{x})$ and is, in fact, a three-dimensional array. This will exist in the sense of Frechet if a linear operator $\mathbf{f_{xx}}(\mathbf{x})$ exists, such that

$$\lim_{h \to o} \frac{\|\mathbf{f_x}(\mathbf{x} + \mathbf{h}) - \mathbf{f_x}(\mathbf{x}) - \mathbf{f_{xx}}(\mathbf{x})\mathbf{h}\|}{\|\mathbf{h}\|} = 0, \tag{B. 25}$$

where $\mathbf{f_{xx}}(x)h$ is a matrix whose ith column is formed from $(\partial \mathbf{f_x}(x)h/\partial x_i)$. Clearly, $\mathbf{f_{xx}}(x)$ will exist if and only if all second partial derivatives (that is, $(\partial^2 f_l/\partial x_i \partial x_j)$, $l = 1, 2, \ldots, m$; $i = 1, 2, \ldots, n$; $j = 1, 2, \ldots, n$) exist and are continuous.

Definition B. 15. Gradient

Let L(x) be a function from R_n into R_1. Then, the derivative of L with respect to the vector **x** is a vector whose kth coordinate is the partial derivative of L(**x**) with respect to the kth coordinate of **x**, and it is called the gradient of L(**x**) with respect to **x**. The gradient of L(**x**) with respect to **x** will be denoted by $L_x(x)$.

Definition B. 16. Vector differential equation

An equation of the form

$$\dot{\mathbf{x}}(t) = \mathbf{f}(\mathbf{x}(t), \mathbf{u}(t), t), \qquad \mathbf{x} \in R_n, \qquad u \in R_m, \qquad t \in R_1 \tag{B.26}$$

is called a first-order vector differential equation, and **x** is called the state.

Theorem B. 3[1]

Let $f(\mathbf{x}, u, t) = \mathbf{A}(t)\mathbf{x} + B(t)\mathbf{u}$, where **A** and B are matrixes whose elements are continuous functions on R_1. Let $\mathbf{\Phi}(t, t_0)$ be the $n \times n$ matrix which is the solution of the matrix differential equation

$$\frac{d}{dt}\mathbf{\Phi}(t, t_0) = A(t)\Phi(t, t_0), \qquad \Phi(t_0, t_0) = I. \tag{B.27}$$

Then, the solution of the equation

$$\dot{\mathbf{x}}(t) = \mathbf{A}(t)\mathbf{x}(t) + \mathbf{B}(t)\mathbf{u}(t), \qquad \mathbf{x}(t_0) = \mathbf{x}_0, \tag{B.28}$$

where $\mathbf{u}(t)$ is defined over the interval $(t_0, t]$, is

$$\mathbf{x}(t; \mathbf{x}_0, t_0; \mathbf{u}_{(t_0, t]}) = \mathbf{\Phi}(t, t_0)\mathbf{x}_0 + \int_{t_0}^{t} \mathbf{\Phi}(t, \xi)B(\xi)\mathbf{u}(\xi)d\xi. \tag{B.29}$$

Remarks:

Some properties of $\mathbf{\Phi}(t, t_0)$ are listed as follows:

1. If $\mathbf{A}(t)$ is constant, then

$$\mathbf{\Phi}(t, t_0) = e^{\mathbf{A}(t-t_0)}. \tag{B.30}$$

2. If $\mathbf{A}(t) \int_{t_0}^{t} \mathbf{A}(\tau)d\tau = [\int_{t_0}^{t} \mathbf{A}(\tau)d\tau]\, \mathbf{A}(t)$ for all t, then

$$\mathbf{\Phi}(t, t_0) = \exp [\int_{t_0}^{t} A(\tau)d\tau]. \tag{B.31}$$

3. If $\mathbf{A}(t)$ is continuous over every finite interval, then $\mathbf{\Phi}(t, t_0)$ is nonsingular for all finite t.
4. $\mathbf{\Phi}(t_1, t_2)\mathbf{\Phi}(t_2, t_3) = \mathbf{\Phi}(t_1, t_3)$. (B.32)
5. $\mathbf{\Phi}(t, t_0) = \mathbf{X}(t)\mathbf{X}^{-1}(t_0)$, $\mathbf{X}(t)$ nonsingular.
6. $\mathbf{\Phi}(t, t_0)$ is called the state transition matrix or the fundamental matrix.
7. $\mathbf{\Phi}(t, t_0) = \mathbf{X}(t)\mathbf{X}^{-1}(t_0)$, where $\overline{\mathbf{X}}(t)$ satisfies Equation B. 27.

[1] See Reference 24, p. 342.

Appendix C. Some Theorems on Normality

Theorem C. 1

If the plant in Problem Z_1 is linear and constant, then Problem Z_1 is normal if[1] each of the m controllability matrices $\mathbf{C}_\beta$, $\beta = 1, 2, \ldots, m$ (see Definition 1.7) is nonsingular.

Proof:

Suppose that $\mathbf{C}_\beta$ are all nonsingular and that the Problem Z_1 is singular. In Problem Z_1, we have

$$H = \mathbf{p}_0 + \langle \mathbf{p}, \mathbf{Ax} \rangle + \sum_{\beta=1}^{m} \langle \mathbf{p}, \mathbf{b}_\beta \rangle w_\beta, \tag{C. 1}$$

and

$$\dot{\mathbf{p}}(t) = -\mathbf{A}'\mathbf{p}(t), \qquad \mathbf{p}(0) = \pi; \tag{C. 2}$$

that is,

$$\mathbf{p}(t) = e^{-\mathbf{A}'t}\pi. \tag{C. 3}$$

Then, we see that

$$w_\beta(t) = -\operatorname{sgn}\{\langle \mathbf{p}(t), \mathbf{b}_\beta \rangle\}, \qquad \beta = 1, 2, \ldots, m. \tag{C. 4}$$

But $w_\beta(t)$ is well defined unless for some β we have $\langle \mathbf{p}(t), \mathbf{b}_\beta \rangle$ identically zero over a nonzero interval for some $\mathbf{p}(t)$ satisfying Equation C. 3. In other words, there exists a $\pi \neq 0$, such that $\langle \boldsymbol{\pi}, e^{-\mathbf{A}t}\mathbf{b}_\beta \rangle$ is iden-

[1] The condition is also necessary [3].

tically zero. Let this $\pi = \xi$ and the $\beta = 1$. Then, taking derivatives, we must have

$$\langle \xi, \mathbf{A}^k e^{-\mathbf{A}t} \mathbf{b}_1 \rangle = 0, \quad k = 0, 1, \ldots, n-1. \tag{C.5}$$

Equations 1.37 can be rewritten in the form

$$\mathbf{C}'_1 e^{-\mathbf{A}'t} \xi = 0. \tag{C.6}$$

But the matrix $\mathbf{C}'_1 e^{-\mathbf{A}'t}$ is nonsingular. Hence, $\xi = 0$ is the only solution to Equation C.6, and thus Z_1 cannot be singular.

Theorem C.2

If the plant in Problem Z_2 is linear and constant, then Z_2 is normal if <u>one</u> of the controllability matrices $\mathbf{C}_\beta, \beta = 1, 2, \ldots, m$ is nonsingular.

Proof:

In this case, we have

$$H = p_0 + \langle \mathbf{p}, \mathbf{A}\mathbf{x} \rangle + \langle \mathbf{p}, \beta \mathbf{w} \rangle, \tag{C.7}$$

from which we obtain

$$\mathbf{w}(t) = - \frac{\beta' \mathbf{p}(t)}{\|\beta' \mathbf{p}(t)\|}. \tag{C.8}$$

Here $\mathbf{w}(t)$ is well defined (for the same arguments as used in Theorem 1.3) for all $\mathbf{p}(t)$ unless <u>all</u> of the $\mathbf{C}_\beta$ are singular.

Theorem C.3

If the plant in Problem Z_3 is linear and constant, then Z_3 is normal if both the controllability matrix $\mathbf{C}$ and the system matrix $\mathbf{A}$ are nonsingular.

Proof:

Suppose that $\mathbf{C}$ and $\mathbf{A}$ are nonsingular and that Z_3 is singular. For Problem Z_3 with this plant, we have

$$H = p_0 |w| + \langle \mathbf{p}, \mathbf{A}\mathbf{x} \rangle + \langle \mathbf{p}, \mathbf{b} \rangle w. \tag{C.9}$$

If $p_0 = 0$, we have the Hamiltonian of Problem Z_1. Hence, if $t_1 > t_1^*$, then $p_0 = 1$. Under this assumption, we see that

$$w(t) = - \operatorname{dez} \{\langle \mathbf{p}(t), \mathbf{b} \rangle\}, \tag{C.10}$$

where $\mathbf{p}(t)$ satisfies Equation 1.34, and

$$\text{dez}\,\{\alpha\} = \begin{cases} +1, & \alpha > 1, \\ -1, & \alpha < -1, \\ 0, & |\alpha| > 1, \\ ?, & |\alpha| = 1. \end{cases} \tag{C.11}$$

If the problem is normal, there exists a vector $\pi = \xi$, such that $\langle \xi, e^{-\mathbf{A}t}\mathbf{b}\rangle$ is identically plus or minus 1 over some nonzero interval in time. Or, taking derivatives, we have

$$\mathbf{C}'\mathbf{A}'e^{-\mathbf{A}t}\xi = 0, \tag{C.12}$$

which again can occur only for $\xi = 0$.

Theorem C.4

If the plant in Problem Z_4 is linear and the fixed time $T > T^*$, the Problem Z_4 is normal.

Proof:

$$H = p_0 \frac{1}{\rho} |w|^{\rho} + \langle \mathbf{p}, \mathbf{A}(t)\mathbf{x}\rangle + \langle \mathbf{p}, \mathbf{b}(t)\rangle w. \tag{C.13}$$

Since $T > T^*$,

$$p_0 = 1,$$

and

$$w(t) = -|\langle \mathbf{p}(t), \mathbf{b}(t)\rangle|^{1/(\rho-1)} \operatorname{sgn}\{\langle \mathbf{p}(t), \mathbf{b}(t)\rangle\}. \tag{C.14}$$

Equation 1.44 is well defined at $\langle \mathbf{p}(t), \mathbf{b}(t)\rangle = 0$.

The remainder of this appendix contains a more general discussion the question of normality. In particular, the question of the normality of Problem Z_5 is treated in some detail.

The Hamiltonian associated with Problem Z_5 is

$$H(\mathbf{x}, \mathbf{u}, \mathbf{p}, p_0, t) = g(\|\mathbf{u}\|) + \langle \mathbf{p}, \mathbf{B}(t)\mathbf{u}\rangle + \langle \mathbf{p}, \mathbf{A}(t)\mathbf{x}\rangle. \tag{C.15}$$

Since the last term in Equation C.15 is independent of $\mathbf{u}$, it can be ignored, and since we are interested in minimizing H at each instant of time t, we can consider only the minima of the function

$$h(\mathbf{u}) = g(\|\mathbf{u}\|) + \langle \mathbf{c}, \mathbf{u}\rangle, \tag{C.16}$$

where $\mathbf{c}(t) = \mathbf{B}'(t)\mathbf{p}(t)$, $\mathbf{p}(t)$ admissible. (Note that $\mathbf{c}$ is an n-vector.)

It is convenient to define two simple sets in R_m. Let

$$N(\rho) = (\mathbf{u} : \mathbf{u} \in R_m, \|\mathbf{u}\| \leqslant \rho), \quad \rho \geqslant 0, \tag{C.17}$$

and

$$R(\mathbf{u}_1, \mathbf{e}) = (\mathbf{u} : \mathbf{u} \in R_m, \mathbf{u} = \mathbf{u}_1 + \lambda\mathbf{e}, \quad 0 \leq \lambda < \infty, \mathbf{e} \in R_m, \quad \|\mathbf{e}\| = 1). \qquad (C.18)$$

In words, N(1) is a closed neighborhood of the origin of R_m (a closed hypersphere or ball of radius 1), and $R(\mathbf{u}_1, \mathbf{e})$ is a "ray" from the point $\mathbf{u}_1$, having the direction of $\mathbf{e}$.

Properties of h

P1. The inner product $\langle \mathbf{c}, \mathbf{e}\rangle$, $\mathbf{e} \in N(1)$, is minimized on N(1) at the point $\mathbf{e}^{**} \in \partial N(1)$, whose support hyperplane has a normal, outward from N(1), whose direction is that of $-\mathbf{c}$. Furthermore, $\mathbf{e}^{**}$ is unique unless $\partial N(1)$ contains a line segment that is perpendicular to $\mathbf{c}$.

Proof:

Let

$$P(\mathbf{c}, \beta) = (\mathbf{u} : \langle \mathbf{c}, \beta\,(\mathbf{c} - \mathbf{u})\rangle = 0, \quad -\infty < \beta < \infty; \qquad (C.19)$$

that is, $P(c, \beta)$ is a hyperplane which is perpendicular to $\mathbf{c}$ and passes through the point βc. If $\mathbf{u} \in P(c, \beta)$, then

$$\langle \mathbf{c}, \mathbf{u}\rangle = \beta\langle \mathbf{c}, \mathbf{c}\rangle \qquad (C.20)$$

a constant. Furthermore, if $\beta_1 < \beta_2$ and $\mathbf{u}_i \in P(c, \beta_i)$, $i = 1, 2$, then $\langle \mathbf{c}, \mathbf{u}_1\rangle < \langle \mathbf{c}, \mathbf{u}_2\rangle$. Since N(1) is convex and compact, there exist two unique numbers β_1 and β_2, with $\beta_2 > \beta_1$, say such that $P(c, \beta_i)$, $i = 1, 2$ are the two unique hyperplanes in R_m having normals in the direction of c and which support N(1). Clearly, if $\mathbf{e}^{**} \in P(\mathbf{c}, \beta_1) \cap \partial N(1)$, then $\mathbf{e}^{**}$ minimizes $\langle \mathbf{c}, \mathbf{e}\rangle$ on N(1). If $\mathbf{e}_m \in \mathbf{P}(\mathbf{c}, \beta_2) \cap \partial N(1)$, then $\mathbf{e}_m$ maximizes $\langle \mathbf{c}, \mathbf{e}\rangle$ on N(1). Finally, $\mathbf{e}^{**}$ is unique, unless $\partial N(1)$ contains a line segment that is also contained in $P(\mathbf{c}, \beta_1)$ (that is, perpendicular to $\mathbf{c}$), in which case $\langle \mathbf{c}, \mathbf{e}\rangle$ is minimized everywhere on that line segment.

P2. If g is strictly convex down, that is, if $g((1 - \lambda)x_1 + \lambda x_2) < (1 - \lambda)g(x_1) + \lambda g(x_2)$, $0 < \lambda < 1$, $x_1 \neq x_2$, and iff (that is, if and only if) $\mathbf{e}^{**}$ is unique (see P1), either there exists a unique vector $\mathbf{u}^{**}$ that minimizes $h(\mathbf{u})$ on R_m, or the minimum of $h(\mathbf{u})$ does not exist, in the sense that $h(\mathbf{u})$ is minimized at the point where $R(\mathbf{0}, \mathbf{e}^{**})$ intersects the closure of R_m.

Proof:

On any ray $R(\mathbf{0}, \mathbf{e})$ from the origin of R_m, we have

$$h(\mathbf{u}) = h(\lambda\mathbf{e}) = g(\lambda) + \lambda\langle \mathbf{c}, \mathbf{e}\rangle. \qquad (C.21)$$

Therefore, if $\mathbf{e}^{**}$ uniquely minimizes $\langle \mathbf{c}, \mathbf{e} \rangle$ on $\partial N(1)$,

$$g(\lambda) + \lambda\langle \mathbf{c}, \mathbf{e}^{**} \rangle < g(\lambda) + \lambda\langle \mathbf{c}, \mathbf{e} \rangle, \tag{C.22}$$

then, it follows that the minimum, if it exists, occurs on the ray $R(\mathbf{0}, \mathbf{e}^{**})$. But h is a function of the scalar λ on $R(\mathbf{0}, \mathbf{e}^{**})$ which is strictly convex down. Now, a function on R_1, which is strictly convex down, attains a unique minimum on either R_1 or on its closure.

Suppose that $\mathbf{e}^{**}$ is not unique. Then, the foregoing will also hold on every ray associated with a vector $\mathbf{e}$ that minimizes $\langle \mathbf{c}, \mathbf{e} \rangle$ on $\partial N(1)$.

P3. Let $\mathbf{e}^{**}$ be a vector that minimizes $\langle \mathbf{c}, \mathbf{e} \rangle$ on $\partial N(1)$. Suppose that the function g has linear pieces (that is, suppose that g is not strictly convex). Then, (1) if the derivative $g(x) = k$, a constant, for $\lambda_1 \leqslant x \leqslant \lambda_2$ and $\langle \mathbf{c}, \mathbf{e}^{**} \rangle = -k$, then $h(\mathbf{u})$ is minimized for $\mathbf{u}^{**} = \lambda\mathbf{e}^{**}, \lambda_1 \leqslant \lambda \leqslant \lambda_2$; (2) if $g_x(x)$ has a maximum value $M < \infty$ (recall that g is assumed to be monotone increasing) and $|\langle \mathbf{c}, \mathbf{e}^{**} \rangle| \geqslant M$, then $h(\mathbf{u})$ is minimized on the closure R_m at $\mathbf{u}^{**} = \lim_{\lambda \to \infty} \lambda\mathbf{e}^{**}$; (3) if $\mathbf{c}$ is perpendicular to a line segment of $\partial N(1)$, (1) and (2) hold for any $\mathbf{e}$ on that line segment. Otherwise, h is minimized by a unique vector $\mathbf{u}^{**}$ in R_m.

Proof:

If $g(\lambda) = -\langle \mathbf{c}, \mathbf{e}^{**} \rangle$ for $\lambda_1 \leqslant \lambda \leqslant \lambda_2$, then $h(\lambda\mathbf{e}^{**})$ is a constant on the interval $[\lambda_1, \lambda_2]$ and is a minimum, since h is convex down. Property P3(1) is then proved. Property P3(2) is simply the case mentioned in Property P2 where the minimum occurs on the closure of R_m. Property P3(3) is obvious, and, since g is monotone increasing and convex down, this is the only case where the minimum can occur on the closure of R_m. It follows that $\mathbf{u}^{**}$ is unique if (1), (2), or (3) does not hold.

The case where $g(\|\mathbf{u}\|) = \|\mathbf{u}\|$ (that is, where g has two linear pieces with slopes ± 1, respectively) is of particular interest, since it includes the so-called minimum-fuel problems. The following two norms have appeared in the literature, and $\partial N(1)$ contains line segments for both of these:

$$\|\mathbf{u}\| = \sum_{i=1}^{m} |u_i|, \tag{C.23}$$

and

$$\|\mathbf{u}\| = \sup\,(|u_i|)_{i=1}{}^{m}. \tag{C.24}$$

In addition, it is sometimes useful to define Ω in terms of norms, that is,

$$\Omega = (\mathbf{u} : \|\mathbf{u}\| \leqslant 1). \tag{C.25}$$

The use of the sup norm in defining Ω corresponds to having m saturating controllers, all of which can act independently (that is, they have their own power supply). The use of Equation C. 23 in Equation C. 25 can be realized for $m = 2$ by using two jets at right angles fed from the same source, such that the total maximum flow from the source, assumed equal to the sum of the magnitudes of the u_i, is fixed. Indeed, in a given application, one might have a combination of both. For example,

$$\Omega = \left(\mathbf{u} : \sum_{i=1}^{k} |u_i| + \sup \left(|u_i|\right)_{i=k+1}{}^{m} \leq 1 \right).$$

It is apparent from Properties P1, P2, and P3 that these "practical" situations might be difficult, since N(1) contains line segments and g is not strictly convex.

Theorem C. 5

If (1) $\Omega = (\mathbf{u}: \|\mathbf{u}\| \leq 1)$, that is, $\Omega = N(1)$, (2) $g_x(x) = k$ on $[\lambda_1, \lambda_2]$ and either $\langle \mathbf{c}, \mathbf{e}^{**} \rangle \neq -k$ or $\lambda \mathbf{e}^{**} \notin N(1)$ or λ in $[\lambda_1, \lambda_2]$, and (3) $\mathbf{c}$ is not perpendicular to a line segment of $\partial N(1)$, then there exists a unique vector $\mathbf{u}^{**}$ that minimizes $h(\mathbf{u})$ on Ω.

Proof:

By definition, $\mathbf{e}^{**}$ minimizes $\langle \mathbf{c}, \mathbf{e} \rangle$ on N(1). But, since $\partial\Omega = \partial N(1)$, $\mathbf{e}^{**}$ also minimizes $\langle \mathbf{c}, \mathbf{e} \rangle$ on $\partial\Omega$, and it also minimizes $h(\mathbf{u})$ on $\partial\Omega$, since $g(\|\mathbf{u}\|)$ is a constant on Ω. Hence, either $\mathbf{u}^{**} = \mathbf{e}^{**}$ or $\mathbf{u}^{*} \in \Omega/\partial\Omega$ (the interior of Ω). The uniqueness therefore follows from Properties P1 through P3 of h.

Theorem C. 6

If g is strictly convex down, if Ω is compact and convex, and if $\mathbf{c}$ is not perpendicular to a line segment of $\partial N(1)$, there exists a unique vector $\mathbf{u}^*$ that minimizes $h(\mathbf{u})$ on Ω.

If Ω is compact and strictly convex, if $-\langle \mathbf{c}, \mathbf{e}^{**} \rangle \neq$ slope of a linear piece of g, and if $\mathbf{c}$ is not perpendicular to a line segment of $\partial N(1)$, there exists a unique vector $\mathbf{u}^*$ that minimizes $h(\mathbf{u})$ on Ω.

Proof:

Let

$$G(\rho) = (\mathbf{u} : h(\mathbf{u}) \leq \rho), \qquad \rho \geq h(\mathbf{u}^{**}), \tag{C. 26}$$

where

$$h(\mathbf{u}^{**}) = \min_{\mathbf{u} \in R_m} (h(\mathbf{u})). \tag{C. 27}$$

The set $G(\rho)$ is obviously compact if $\mathbf{u}^{**} \in R_m$, and if $\mathbf{u}^{**}$ is on the closure of R_m, $G(\rho)$ has a well-defined boundary in R_m. It is now shown that $G(\rho)$ is convex. Suppose that $h(\mathbf{u}_1) = h(\mathbf{u}_2) = \rho$ and suppose that $\mathbf{u} = (1-\lambda)\mathbf{u}_1 + \lambda\mathbf{u}_2, 0 \leq \lambda \leq 1$. Then, we have

$$h(\mathbf{u}) = g(\|(1-\lambda)\mathbf{u}_1 + \lambda\mathbf{u}_2\|) + (1-\lambda)\langle\mathbf{c}, \mathbf{u}_1\rangle + \lambda\langle\mathbf{c}, \mathbf{u}_2\rangle. \tag{C.28}$$

but

$$\|(1-\lambda)\mathbf{u}_1 + \lambda\mathbf{u}_2\| \leq (1-\lambda)\|\mathbf{u}_1\| + \lambda\|\mathbf{u}_2\|, \tag{C.29}$$

and hence

$$g(\|(1-\lambda)\mathbf{u}_1 + \lambda\mathbf{u}_2\|) \leq g((1-\lambda)\|\mathbf{u}_1\| + \lambda\|\mathbf{u}_2\|), \tag{C.30}$$

and the equality holds in Equation C.30 only if the equality holds in Equation C.29, since g is monotone increasing.

Since g is convex down, we see that

$$g((1-\lambda)\|\mathbf{u}_1\| + \lambda\|\mathbf{u}_2\|) \leq (1-\lambda)g(\|\mathbf{u}_1\|) + \lambda g(\|\mathbf{u}_2\|), \tag{C.31}$$

where the equality holds only if $\|\mathbf{u}_1\|$ and $\|\mathbf{u}_2\|$ are in the domain of the same linear piece of g. Then,

$$h((1-\lambda)\mathbf{u}_1 + \lambda\mathbf{u}_2) \leq (1-\lambda)h(\mathbf{u}_1) + \lambda h(\mathbf{u}_2), \tag{C.32}$$

where the equality holds only if both (1) $\|\mathbf{u}_1\|$ and $\|\mathbf{u}_2\|$ are on the same linear piece of g, and (2) $\mathbf{u}_1$ and $\mathbf{u}_2$ satisfy Equation C.29 with the equality. Then, $G(\rho)$ is convex and is strictly convex if (1) or (2) cannot hold. The convexity of $G(\rho)$ is of interest, since $\mathbf{u}^{**}$ is either in Ω or it is not. If $\mathbf{u}^{**}$ is unique (see Properties P2 and P3) and $G(\rho)$ is convex, then $\mathbf{u}^*$ will be unique if Ω is strictly convex. Similarly, $\mathbf{u}^{**}$ will be unique if $G(\beta)$ is strictly convex and Ω is only convex. But $\mathbf{u}^{**}$ is unique unless $g_x(x) = -\langle\mathbf{c}, \mathbf{e}^{**}\rangle = k$ (see Property P3(1)) or $-\mathbf{c}$ is perpendicular to a line segment of $\partial N(1)$ (see Property P3(3)), which proves the theorem.

Theorems C.5 and C.6 show that the problem will be normal when g, N(1), and Ω contain no line segments. However, these conditions are only sufficient. The theorems also state that the problem will be normal if g is strictly convex down and $\mathbf{c}(t)$ can be perpendicular to a line segment of $\partial N(1)$ at most a finite number of times in any interval $(t_0, T]$. Conversely, the problem will also be normal when g is not strictly convex if Ω is strictly convex and $\mathbf{c}(t)$ can be perpendicular to a line segment of $\partial N(1)$, or if $-\langle\mathbf{c}, \mathbf{e}\rangle$ equals the slope of a linear piece of g at most a finite number of times in any interval $(t_0, T]$.

For example, if $\Omega = (\mathbf{u}\colon |u_i| \leq 1, i = 1, 2, \ldots, m)$ and $\mathbf{u} = \left(\sum_{i=1}^{m} (1/\rho)\,|u_i|^\rho\right)^{1/\rho}, 1 \leq \rho < \infty$, the problem will be normal if g is strictly convex.

Theorem C. 5 treats the case where g, N(1), and Ω are only convex, and it is shown that the problem will be normal if $\mathbf{c}(t)$ can be perpendicular to a line segment of $\partial N(1)$, or if $-\langle \mathbf{c}(t), \mathbf{e}\rangle$ equals the slope of a linear piece of g at most a finite number of times in any interval $(t_0, T]$ (that is, the same conditions as formerly) when one chooses $\Omega = N(1)$.

The conditions just deduced can be used to derive sufficient conditions that ensure normality, as can those of Theorems C. 1 and C. 3. For example, suppose that both $\mathbf{A}(t)$ and $\mathbf{B}(t)$ are constant, $\Omega = N(1) = (\mathbf{u}\colon \sum_{i=1}^{m} |u_i| \leqslant 1)$, and $g(x) = |x|$. Theorem C. 5 applies to this case. If the problem is not normal, there exists a vector $\mathbf{e}_1 \neq \mathbf{0}$, such that $-\langle \mathbf{c}(t), \mathbf{e}_1\rangle = 1$ identically on some nonzero subinterval of $(t_0, T]$, or there exists a vector $\mathbf{e}_2 \neq 0$, such that $\langle \mathbf{c}(t), \mathbf{e}_2\rangle = 0$ identically, or both. Now, $\langle \mathbf{c}(t), \mathbf{e}\rangle = \langle \mathbf{p}(t), \mathbf{B}\mathbf{e}\rangle = \langle \boldsymbol{\pi}, e^{-\mathbf{A}t}\mathbf{B}\mathbf{e}\rangle$. Hence, if the problem is not normal, either $\langle \boldsymbol{\pi}, (-1)^k e^{-\mathbf{A}t}\mathbf{A}^k\mathbf{B}\mathbf{e}_1\rangle = 0, k = 1, 2, \ldots$ (that is, taking derivatives with respect to time), or $\langle \boldsymbol{\pi}, (-1)^k e^{-\mathbf{A}t}\mathbf{A}^k\mathbf{B}\mathbf{e}_2\rangle = 0,\ k = 0, 1, 2, \ldots$, or both. Consider the latter. This condition implies that there exists a vector $\boldsymbol{\pi} \neq 0$ in R_n, such that

$$\boldsymbol{\pi}' e^{-\mathbf{A}t}\mathbf{A}[\mathbf{B}\mathbf{e}_2 \,\vdots\, -\mathbf{A}\mathbf{B}\mathbf{e}_2 \,\vdots\, \ldots\ (-1)^n\mathbf{A}^{n-1}\mathbf{B}\mathbf{e}_2] = \mathbf{0}.$$

But this cannot hold if the matrix multiplying $\boldsymbol{\pi}$ is nonsingular for any vector $\mathbf{e}_2 \neq \mathbf{0}$ in R_m. Now, $e^{-\mathbf{A}t}$ is nonsingular. Hence, the problem will be normal if both $\mathbf{A}^{-1}$ exists and the matrix

$$[\mathbf{B}\mathbf{e}_2 \,\vdots\, \mathbf{A}\mathbf{B}\mathbf{e}_2 \,\vdots\, \ldots\ \mathbf{A}^{n-1}\mathbf{B}\mathbf{e}_2] = \left[\sum_{i=1}^{m} \mathbf{b}_i e_{2i} \,\vdots\, \mathbf{A}\sum_{i=1}^{m} \mathbf{b}_i e_{2i} \ \ldots \,\vdots\, \mathbf{A}^{n-1} \sum_{i=1}^{m} \mathbf{b}_i e_{2i}\right]$$

is nonsingular for all vectors $\mathbf{e}_2 \neq 0$ in R_m. Recognizing that the controllability matrix for the ith controller is $\mathbf{C}_i = [\mathbf{b}_i \,\vdots\, \mathbf{A}\mathbf{b}_i \,\vdots\, \ldots\ \mathbf{A}^{n-1}\mathbf{b}_i]$, we see that the condition can be restated as follows: the problem is normal if both the matrix $\mathbf{A}$ and the matrix $\sum_{i=1}^{m} e_{2i}\mathbf{C}_i$ are nonsingular for all vectors $\mathbf{e}_2 \neq 0$ in R_m. This condition is not very useful as it stands, unless we know what properties the matrices $\mathbf{C}_i$ must have so that all linear combinations of them will be nonsingular. This question and that of other conditions ensuring normality are left as subjects for further research. It is noted that if $m = 1$, the condition just found coincides with that of Theorem C. 3 as it should.

Appendix D. Existence of a State-Space Representation in which the Euclidean Norm is a Liapunov Function

In this appendix, it is shown that if all the eigenvalues of the real $n \times n$ matrix $\mathbf{A}$ have negative real parts, where

$$\dot{\mathbf{x}}(t) = \mathbf{A}\mathbf{x}(t) + \mathbf{B}\mathbf{u}(t) \tag{D. 1}$$

then there exists a state-space representation for the plant of Equation D. 1 for which the associated plant matrix $\tilde{\mathbf{A}}$ is negative definite. In other words, given the plant of Equation D. 1 and given that all of the eigenvalues of $\mathbf{A}$ have negative real parts, there exists an $n \times n$ nonsingular real constant matrix $\mathbf{P}$, such that if

$$\mathbf{x} = \mathbf{P}\mathbf{y}, \tag{D. 2}$$

and

$$\tilde{\mathbf{A}} \equiv \mathbf{P}^{-1}\mathbf{A}\mathbf{P}, \tag{D. 3}$$

then

$$\dot{\mathbf{y}}(t) = \tilde{\mathbf{A}}\mathbf{y}(t) + \mathbf{P}^{-1}\mathbf{B}\mathbf{u}(t), \tag{D. 4}$$

and

$$-\langle \mathbf{y}, \tilde{\mathbf{A}}\mathbf{y} \rangle \geqslant 0, \tag{D. 5}$$

and the equality holds only if $\mathbf{y} = \mathbf{0}$ (that is, the matrix $-\tilde{\mathbf{A}}$ is positive definite). As is well known (see, for example, [16] Vol. II), if all the eigenvalues of $\mathbf{A}$ have negative real parts, then corresponding to any negative quadratic form $\langle \mathbf{x}, \mathbf{W}\mathbf{x} \rangle$, there corresponds a positive definite quadratic form (a Liapunov function) $\langle \mathbf{x}, \mathbf{H}'\mathbf{H}\mathbf{x} \rangle$, where $\mathbf{H}$ is $n \times n$ with rank n, such that if

$$\frac{d}{dt}\mathbf{x} = \mathbf{A}\mathbf{x}, \tag{D. 6}$$

then

$$\frac{d}{dt}\langle \mathbf{x}, \mathrm{HH}'\mathbf{x} \rangle = \langle \mathbf{x}, \mathbf{W}\mathbf{x} \rangle. \tag{D. 7}$$

Let

$$\mathbf{P} = \mathbf{H}', \tag{D. 8}$$

and $d/dt\langle \mathbf{y}, \mathrm{y} \rangle$ is negative definite. But

$$\frac{d}{dt}\langle \mathbf{y}, \mathrm{y} \rangle = \langle \mathbf{y}, (\tilde{\mathbf{A}}' + \tilde{\mathbf{A}})\mathrm{y} \rangle = 2\langle \mathrm{y}, \tilde{\mathbf{A}}\mathbf{y} \rangle, \tag{D. 9}$$

or $-\tilde{\mathbf{A}}$ is positive definite.

Q.E.D.

Appendix E. Hyperspherical Coordinates

In this appendix, it is shown that Properties 1 and 2 of $h_\theta(\theta)$ hold (see Definition 4. 11). The jth column of $\mathbf{h}_\theta(\theta)$ may be written

$$
\mathbf{h}_\theta(\theta) = (\text{jth column}) \begin{bmatrix} 0 \\ \vdots \\ 0 \\ \hline -\prod_{l=1}^{j} \sin\theta_l \\ \hline \vdots \\ \cos\theta_j \cos\theta_i \prod_{\substack{l=1 \\ \neq j}}^{i-1} \sin\theta_l \\ \vdots \\ \hline \cos\theta_j \prod_{\substack{l=1 \\ \neq j}}^{n-1} \sin\theta_l \end{bmatrix} \begin{matrix} \left.\right\} j-1 \text{ zeros}, \\ \\ \left.\right\} i = j, \\ \\ \left.\right\} i = j+1,\ j+2 \ldots,\ n-1, \\ \\ \left.\right\} i = n. \end{matrix} \qquad (\text{E. }1)
$$

Let $\mathbf{h}_k$ and $\mathbf{h}_j$ denote the kth and the jth columns, respectively, of $\mathbf{h}_\theta(\theta)$. Then, we have

$$\langle \mathbf{h}_k, \mathbf{h}_j \rangle = \left(-\prod_{l=1}^{j} \sin \theta)(\cos \theta_k \cos \theta_j \prod_{\substack{m=1\\ \neq k}}^{j-1} \sin \theta_m\right)$$

$$+ \sum_{\lambda=j+1}^{n-1} \left(\cos \theta_j \cos \theta_\lambda \prod_{\substack{l=1\\ \neq j}}^{\lambda-1} \sin \theta_l)(\cos \theta_k \cos \theta_\lambda \prod_{\substack{m=1\\ \neq k}}^{\lambda-1} \sin \theta_m\right)$$

$$+ \sin \theta_j \cos \theta_k \left(\prod_{\substack{l=1\\ \neq j}}^{n-1} \sin \theta_l\right)\left(\prod_{\substack{m=1\\ \neq k}}^{n-1} \sin \theta_m\right). \tag{E. 2}$$

Factoring out the largest common divisor in Equation E. 2, we have

$$\langle \mathbf{h}_k, \mathbf{h}_j \rangle = \cos \theta_k \cos \theta_j \sin \theta_k \sin \theta_j \prod_{\substack{m=1\\ \neq k}}^{j-1} \sin \theta_m$$

$$\times \left\{-1 + \cos^2 \theta_{j+1} + \sum_{\lambda=j+1}^{n-2} \cos^2 \theta_{\lambda+1} \prod_{m=j+1}^{\lambda} \sin^2 \theta_m\right.$$

$$\left. + \prod_{m=j+1}^{n-1} \sin^2 \theta_m\right\}. \tag{E. 3}$$

The terms inside the braces in Equation E. 3 sum to zero by repeated use of the identity

$$\sin^2 \theta + \cos^2 \theta = 1. \tag{E. 4}$$

Hence, the columns of $\mathbf{h}_\theta(\theta)$ are mutually orthogonal. The fact that $\mathbf{h}(\theta)$ is also orthogonal to the $n - 1$ columns of $\mathbf{h}_\theta(\theta)$ follows similarly.

Appendix F. A Program for Computing Time-Optimal Controls

F.1 General Description

The experimental results of Section 8.4 were obtained using the program described in this appendix. This program calculates the time-optimal control that steers the state of a given linear constant plant from a given initial state $\mathbf{x}_0$ to a prescribed hyperspherical target set of radius r, centered at the origin of the state space (that is, it is assumed in the program that $\mathbf{Q} = \mathbf{I}$. See Definition 1.11). The initial guess $(\mathbf{v}_0, T_0)$ can be read in as data, or it can be calculated using the initial-guess procedure described in Section 4.6. This program can be used to find time-optimal controls if (1) the matrix $-\mathbf{A}$ is positive definite, (2) the state-space representation used is such that $\mathbf{A}$ is in its real Jordan form, (3) complex eigenvalues are distinct, (4) real eigenvalues have a multiplicity of at most two, and (5) the plant has a single input. Condition (1) is required by the theoretical aspects of the algorithm, whereas all other requirements can be relaxed by suitably modifying the programs.

All of the subroutines except CGUESS and CTIMES can be used for any plant satisfying the five conditions first specified. CGUESS is used either to read in or to calculate the initial guess. Since the method of problem separation for calculating bounds on the time is strongly dependent on the plant type, this subroutine must be rewritten if the structure of $\mathbf{A}$ is changed. Two subroutines CGUESS are included here; the one for plants having distinct eigenvalues, and the other for plants having two pairs of complex conjugate eigenvalues. Subroutine CTIMES is used to find the switch times $\tau_j(\mathbf{v}, T)$. This is carried out by searching in $(0, T)$ for the zeros of $\langle \mathbf{v}, e^{\mathbf{A}\tau}\mathbf{b} \rangle$, and since this is quite the most time-consuming part of the program, the subroutine was written to minimize the time spent evaluating $\langle \mathbf{v}, e^{\mathbf{A}\tau}\mathbf{b} \rangle$. Hence, one must rewrite a few of the statements in CTIMES if the structure of $\mathbf{A}$ is changed.

F.2 Subroutines Used in All Programs

The following subroutines are used in at least two of the four programs presented in this monograph.

Function CSPROD(X, Y, N)

This function calculates the scalar product of the two N-vectors X and Y, and the equation

$$Z = CSPROD(X, Y, N) \tag{F.1}$$

means

$$Z = \langle X, Y \rangle. \tag{F.2}$$

```
      FUNCTION CSPROD(X,Y,N)
      DIMENSION X(5),Y(5)
       CSPROD=X(1)*Y(1)
      DO 1 I=2,N
1     CSPROD=CSPROD+X(I)*Y(I)
      RETURN
      END
```

Function ANGLE(Y, X)

ANGLE calculates the polar angle of the point (X, Y) in the X ~ Y plane, where the angle is measured positive counterclockwise from the X-axis and

$$-\pi < ANGLE(Y, X) \leq \pi. \tag{F.3}$$

```
      FUNCTION ANGLE(G1,G2)
      PI=3.1415927
      IF(ABSF(G2)-10.E-20)80,80,81
81     IF(G2)84,80,85
84    IF(G1)82,82,83
82    ANGLE=-PI+ATANF(G1/G2)
      GO TO 88
83    ANGLE=PI+ATANF(G1/G2)
      GO TO 88
85    ANGLE=ATANF(G1/G2)
      GO TO 88
80    IF(G1)86,86,87
86    ANGLE=-PI/2.0-ATANF(G2/G1)
      GO TO 88
87    ANGLE=PI/2.0-ATANF(G2/G1)
88    RETURN
      END
```

Subroutine CMLTV(A, N, M, B, X)

CMLTV multiplies the N × M matrix A into the M-vector B to obtain the N-vector X; that is,

$$X = AB. \tag{F.4}$$

```
      SUBROUTINE CMLTV(A,N,M,B,X)
      DIMENSION A(5,5),B(5),X(5)
      DO 190I=1,N
      X(I)=0.0
      DO190J=1,M
190   X(I)=A(I,J)*B(J)+X(I)
      RETURN
      END
```

Function CACOSF(X)

The function CACOSF is used to find inverse cosines; that is, $\text{CACOSF}(X) = \cos^{-1}(X)$.

```
      FUNCTION CACOSF(X)
      IF(ABSF(X)-1.0E-20)1,1,2
1     CACOSF=1.5707964
      RETURN
2     A=SQRTF(ABSF(1.-X*X))/X
      IF(X)3,1,4
4     CACOSF=ATANF(A)
      RETURN
3     CACOSF=3.1415927+ATANF(A)
      RETURN
      END
```

Subroutine CALANG(XG, ANG, N)

Given an N-vector XG and a radius (ANG(N) in the program), CALANG finds an $(n - 1)$-vector of angles (ANG(I), I = 1, 2, ..., N − 1) such that (Equation 4.73)

$$XG = ANG(N)\mathbf{h}(ANG(1), ANG(2), \ldots, ANG(N-1)), \tag{F.5}$$

where it is assumed that $ANG(N) = \|XG\|$. At a pole (Definition 4.11) the angles that are indeterminate are chosen arbitrarily to be zero.

```
      SUBROUTINE CALANG(XG,ANG,N)
C     TO FIND THE HYPERSPHERICAL ANGLES GIVEN XG
      DIMENSION XG(5),ANG(5)
      K1=N-1
      PI=3.1415927
      K3=K1-1
      IF(N-2)17,17,18
17    ANG(1)=ANGLE(XG(N),XG(K1))
      RETURN
18    IF(ABSF(XG(1))-ANG(N)+1.E-20)2,1,1
1     DO 3 I=1,K1
3     ANG(I)=0.
14    RETURN
2     ANG(1)=CACOSF(XG(1)/ANG(N))
```

```
19      DO 6 J=2,K1
        IF(ABSF(XG(J))-1.E-19)7,7,8
7       K2=J-1
        DOOP=ANG(N)
        DO 5 I=1,K2
5       DOOP=DOOP*SINF(ANG(I))
        IF(ABSF(DOOP)-1.E-10)9,9,10
10      IF(J-K1)11,12,6
12      ANG(K1)=ANGLE(XG(N),XG(K1))
        GO TO 6
9       DO 13 I=J,K1
13      ANG(J)=0.
        GO TO 14
11      DO 15 I=J,K3
15      ANG(J)=PI/2.
        J=K3
        GO TO 6
8       IF(J-K1)16,12,6
16      K2=J-1
        DOOP=ANG(N)
        DO 4 I=1,K2
4       DOOP=DOOP*SINF(ANG(I))
        ANG(J)=CACOSF(XG(J)/DOOP)
6       CONTINUE
        RETURN
        END
```

Subroutine CJACO(ANG, N, TEMP)

Given a set of N − 1 angles ANG(I), I = 1, 2, ..., N − 1, and the radius ANG(N), CJACO determines the N × (N − 1) matrix (Equation 4.46)

$$\text{TEMP(I, J)} = \text{ANG(N)}\mathbf{h}_\theta(\text{ANG(1), ANG(2), } \ldots, \text{ ANG(N} - 1)).$$

```
        SUBROUTINE CJACO(ANG,N,TEMP)
        DIMENSION ANG(5),TEMP(5,5),FUN(5,5),CS(5),SN(5)
C       FORMATION OF JACOBIAN OF SPHERICAL COORDINATES
        M=N
        K1=N-1
        DO 1 I=1,K1
        CS(I)=COSF(ANG(I))
 1      SN(I)=SINF(ANG(I))
 223    DO 228 I=1,M
        K3=M+1-I
        DO 228 J=1,K3
 228    FUN(I,J)=ANG(M)
        IF(M-3)243,244,244
 244    DO229I=3,M
        K3=M+2-I
        DO 229 J=K3,K1
 229    FUN(I,J)=0.0
 243    DO 230 J=2,M
        K5=M-J+1
        DO 245 I=1,K5
 245    FUN(J,K5)=FUN(J,K5)*SN(I)
```

```
230   FUN(J,K5)=-FUN(J,K5)
      DO 231 I=1,K1
      FUN(1,I)=FUN(1,I)*CS(I)
      DO 231 J=1,K1
      IF(J-I)232,231,232
232   FUN(1,I)=FUN(1,I)*SN(J)
231   CONTINUE
      IF(M-3)510,247,247
247   DO 233 I=2,K1
      K7=M-I
      DO 233 J=1,K7
      L=M-I+1
      FUN(I,J)=FUN(I,J)*CS(J)*CS(L)
      DO 233 K = 1,K7
      IF(K-J)234,233,234
234   FUN(I,J)=FUN(I,J)*SN(K)
233   CONTINUE
510   DO 500I=1,M
      DO500J=1,K1
      L=M-I+1
500   TEMP(L,J)=FUN(I,J)
      RETURN
      END
```

Subroutine CFUND(A, T, FUN, N)

Given **A** and T, where it is assumed that **A** satisfies the conditions listed in the general description of the program, CFUND calculates the N × N transition matrix $e^{\mathbf{A}T}$; that is,

$$(\mathrm{FUN}(I, J), I = 1, 2, \ldots, N, J = 1, 2, \ldots, N) = e^{\mathbf{A}T}. \qquad (F.6)$$

```
      SUBROUTINE CFUND(A,T,FUN,N)
      DIMENSION A(5,5),FUN(5,5)
      DO 1 I=1,N
      DO 1 J=1,N
1     FUN(I,J)=0.0
      DO 2 I=1,N
2     FUN(I,I)=EXPF(A(I,I)*T)
      DO 3 I=2,N
      IF(A(I,I)-A(I-1,I-1))3,4,3
4     IF(A(I-1,I)+A(I,I-1))7,5,6
5     FUN(I-1,I)=SINF(A(I-1,I)*T)*FUN(I,I)
      FUN(I,I)=FUN(I,I)*COSF(A(I-1,I)*T)
      FUN(I-1,I-1)=FUN(I,I)
      FUN(I,I-1)=-FUN(I-1,I)
      GO TO 3
6     FUN(I-1,I)=T*FUN(I,I)
      GO TO 3
7     TYPE 8
8     FORMAT(16H TROUBLE IN FUND)
      CALL EXIT
3     CONTINUE
      RETURN
      END
```

Subroutine CLINEQ(NLIN, NLP)

CLINEQ solves a matrix equation of the form AX = B, where A is an $n \times n$ matrix, and X and B have n rows and m columns, for the unknown matrix X. The array EH(I, J) must have dimension $n \times n + m$ or larger, and B is stored in its last m columns. We define NLIN = n and NLP = n + m, and the solution X replaces B in the last m columns of EH.

```
      SUBROUTINE CLINEQ(NLIN,NLP)
      COMMON N,T,CONT,KOUNT,M,A(5,5),B(5,2),Q(5,5),XO(5),Z(5),D(5,5),X1(
     15),XH(5),XG(5),KONT,XJ,YJ,DLA,AL,NS(2),TS(2,40),DI(2,40),IC(2),ENM
     2,DENM,X11(5),AI(5,5),Y(5),EH(5,10)
      NL1=NLIN-1
      NL2=NLIN+1
      DO 9609 IL=1,NL1
      ILM=IL
      IL1=IL+1
      BAX=EH(IL,IL)
      DO 9601 JL=IL1,NLIN
      IF(ABS(EH(JL,IL))-ABS(BAX))9601,9601,9602
 9602 BAX=EH(JL,IL)
      ILM=JL
 9601 CONTINUE
      IF(BAX)9603,9604,9603
 9604 RETURN
 9603 IF(ILM-IL) 9605,9606,9605
 9605 DO 9607 JL=IL,NLP
      TEMP=EH(IL,JL)
      EH(IL,JL)=EH(ILM,JL)
 9607 EH(ILM,JL)=TEMP
 9606 TEMP=1./BAX
      DO 9608 JL=IL1,NLP
 9608 EH(IL,JL)=EH(IL,JL)*TEMP
      DO 9609 ILM=IL1,NLIN
      TEMP=EH(ILM,IL)
      DO 9609 JL = IL1,NLP
 9609 EH(ILM,JL)=EH(ILM,JL)-TEMP*EH(IL,JL)
      BAX=EH(NLIN,NLIN)
      IF(BAX)9610,9604,9610
 9610 DO 9612 IL=NL2,NLP
 9612  EH(NLIN,IL)=EH(NLIN,IL)/BAX
      DO 9611 IL=1,NL1
      IL1=NL2-IL
      DO 9611 JL=IL1,NLIN
      DO 9611 ILM=NL2,NLP
 9611 EH(IL1-1,ILM)=EH(IL1-1,ILM)-EH(JL,ILM)*EH(IL1-1,JL)
      RETURN
      END
```

Subroutines CSPV, CSTV, and CVPV

These subroutines are considered self-explanatory and are simply listed here:

```
      SUBROUTINE CSPV(X,Y,A,N,B)
      DIMENSION X(5),Y(5)
      DO 1 I=1,N
1     X(I)=A+Y(I)*B
      RETURN
      END

      SUBROUTINE CSTV(X,Y,A,N)
      DIMENSION X(5),Y(5)
      DO 1 I=1,N
1     X(I)=A*Y(I)
      RETURN
      END

      SUBROUTINE CVPV(X,Y,Z,N)
      DIMENSION X(5),Y(5),Z(5)
      DO 1 I=1,N
1     Z(I)=X(I)+Y(I)*A
      RETURN
      END
```

Subroutine MAIN

MAIN exercises over-all control of the program. The control variables ITYPE [indicates the type of initial-guess procedure being used (see CGUESS)], KONT (an upper limit on the number of iterations), N (the order of the plant), and CONT (if the error becomes less than CONT the iteration stops). The matrix **A** is read in, placed in the first n columns of EH which is then used in CLINEQ to find AI = $\mathbf{A}^{-1}$. The vector **b** (C(I), I = 1, 2, ..., N in the program) the radius r (ANG(N) in the program), and the initial state $\mathbf{x}_0$ (XO(I), i = 1,2, ..., N) are read in as input data. CGUESS is then called, so that an initial guess can be made, CTIMES is then used to calculate the switch times (TS(I)), $u(0^+)$ (D in the program), the number of switchings NS from the given guess XG(I), and T. Also CALXH uses the output of CTIMES and the current guess (that is, XG and T) to calculate $\mathbf{M}(\mathbf{v}, T)$ (XH(I)), $\mathbf{x}_0 - \mathbf{M}(\mathbf{v}, T)$ (E(I)), and $\mathbf{x}_0 - \mathbf{M}(\mathbf{v}, T)$ (ENM). The appropriate data are then printed by subroutine CPRNT. The error and current number of iterations (KOUNT) are then checked, and if the former is not too small and the latter is not too large, CALG is called to calculate the matrix $\mathbf{G}(\mathbf{v}, T)$ and place it in EH for inversion in CLINEQ. Also $\Delta\theta$ (DW(I), I = 1, 2, ..., N − 1) and ΔT (DW(N)) are calculated as per Section 4.4. Subroutine CALNG is then used to obtain the new error, etc., KOUNT is then indexed, the current output is printed by CPRNT, and the cycle is repeated.

The program variables can be associated with the notation of Section 4.4 as follows:

N	n	DW(N)	$\Delta\hat{T}_k$
A(I, J)	$\mathbf{A}$	XG(I)	$\mathbf{v}_k$
C(I)	$\mathbf{b}$	E(I)	$\mathbf{e}_k$
H(I, J)	$\mathbf{h}_\theta(\theta)$	TS(I)	$\tau_i(\mathbf{v}_k, T_k)$
T	T_k	NS	$\mu(\mathbf{v}_k, T_k)$
XO(I)	$\mathbf{x}_0$	ENM	$\Vert\mathbf{e}_k\Vert$
ANG(I)	θ_k	D	$u(=0^+)$
ANG(N)	r	XH(I)	$\mathbf{x}_k$
K1	$n-1$	S	γ_k
AI(I, J)	$\mathbf{A}^{-1}$	KOUNT	k
DW(I)	$\Delta\hat{\theta}_k$		

```
C     TIME OPTIMAL SOLUTION EXPERIMENT NO1 J PLANT
        DIMENSIONXO(5),A(5,5),C(5),ANG(5),EH(5,10),DW(5),XG(5),XH(5),E(5),
       1TS(20),H(5,5),AI(5,5)
        COMMON A,C,XO,CONT,KONT,N,KOUNT,PI,ITYPE,ANG,K1,ITIMES,H,EH,AI
 222    PI=3.1415927
 8      READ 1000,ITYPE,KONT,N,CONT
        K1=N-1
        DO 50 I=1,N
 50     READ 1011,(A(I,J),J=1,N)
        DO 31I=1,N
        DO 31J=1,N
        L=J+N
        EH(I,J)=A(I,J)
 31     EH(I,L)=0.
        DO 41I=1,N
        K=I+N
 41     EH(I,K)=1.
        CALL CLINEQ(EH,N,2*N,M)
        DO 42 I=1,N
        DO 42 J=1,N
        K=J+N
 42     AI(I,J)=EH(I,K)
        READ 1011,(C(I),I=1,N),ANG(N)
 14     READ 1011,(XO(I),I=1,N)
        CALL CGUESS(XG,T)
        S=0.
        KOUNT=0
 4      CALL CTIMES(XG,T,D,NS,TS)
 5      CALL CALXH(XG,NS,T,D,TS,XH,ENM,E)
 30     CALL CPRNT(NS,ENM,T,XG,TS,E,D,S,XH)
        IF(ENM-CONT)8,8,9
 9      IF(KOUNT-KONT)17,8,8
 17     CALL CALG(NS,T,TS,XG,D,XH)
        CALL CLINEQ(EH,N,2*N,M)
        DO 22 I=1,N
        DW(I)=0.
```

```
      DO 22 J=1,N
      K=J+N
22    DW(I)=EH(I,K)*E(J)+DW(I)
      IF(ABSF(DW(N))-.1*T)25,25,300
300   S=.1*T/ABSF(DW(N))
25    CALL CALNG(S,T,XG,NS,D,TS,DW,ENM,E,XH)
      KOUNT=KOUNT+1
      GO TO 30
1011  FORMAT(8F10.5)
1000  FORMAT(3I10,E10.1)
      END
```

Subroutine CGUESS(XG, T)

CGUESS is used for making the initial guess. The input variable (see MAIN) ITYPE controls how this is done.

ITYPE = 1: The initial guess is made using the method of problem separation and the bounds on T outlined in Section 4.4. The statements between and including statements 13 and 10 are written for the case, where **A** has two pairs of complex conjugate eigenvalues and $n = 4$.

ITYPE = 2: The initial guess is read in from data cards.

ITYPE = 3: The initial state $\mathbf{x}_0$ is typed in by the computer operator, and the initial guess is made as for ITYPE = 1.

ITYPE = 4: The initial guess is typed in by the computer operator.

```
      SUBROUTINE CGUESS(XG,T)
      DIMENSIONA(5,5),C(5),XO(5),ANG(5),XG(5),EH(5,10),H(5,5),AI(5,5)
      COMMON A,C,XO,CONT,KONT,N,KOUNT,PI,ITYPE,ANG,K1,ITIMES,H,EH,AI
      GO TO(13,15,24,20),ITYPE
24    PRINT 21
21    FORMAT(14H TYPE XO F10.5)
1012  FORMAT(F10.5)
      ACCEPT 1012,(XO(I),I=1,N)
      GO TO 13
20    PRINT 1011
1011  FORMAT(16H TYPE XG,T,F10.5)
      ACCEPT 1012,(XG(I),I=1,N),T
      GO TO 19
15    READ1013,T,(XG(I),I=1,N)
1013  FORMAT(6E10.4)
      GO TO 19
13    RJ=ANG(N)*ANG(N)
      XONM1=XO(1)*XO(1)+XO(2)*XO(2)
      XONM2=XO(3)*XO(3)+XO(4)*XO(4)
      CNM1=C(1)*C(1)+C(2)*C(2)
      CNM2=C(3)*C(3)+C(4)*C(4)
      D1=CNM1/(2.*ABSF(A(1,1)))
      D2=CNM2/(2.*ABSF(A(3,3)))
      T=LOGF((D1+XONM1)/(D1+RJ))/(2.*ABSF(A(1,1)))
      T1=LOGF((D2+XONM2)/(D2+RJ))/(2.*ABSF(A(3,3)))
      IF(T1-T)9,9,10
```

```
10    T=T1
9     CALL CFUND(A,T,H,N)
      CALL CMLTV(H,N,N,XO,XG)
 19   DUM=ANG(N)/SQRTF(CSPROD(XG,XG,N))
      CALL CSTV(XG,XG,DUM,N)
      RETURN
      END
```

Subroutine CTIMES(X, T, D, NS, TS)

This subroutine is also written for n = 4, and the eigenvalues are all complex. If the nature of the eigenvalues changes, it is necessary to alter the first eight statements and statement 28. Also NS, D, and the TS(I) are calculated here. First D is calculated, and the switch times are found by searching the positive real line in the interval (0, T) for the zeros of $\langle \mathbf{v}, e^{At}\mathbf{b}\rangle$. The search is a simple one of incrementing the time t (see the statement after 34) and checking for a sign change (statement 61). The time increment is increased after each trial, since the function **M** is more sensitive to switch-time accuracy near t = 0. When a sign change is detected, control passes to statement 30, and a further search (by continuously halving the interval) is carried out to determine that particular switch time more accurately. This subroutine consumes by far the greatest portion of the computing time and may well prove to be faster if carried out on an analogue computer.

```
      SUBROUTINE CTIMES(X,T,D,NS,TS)
      DIMENSIONA(5,5),C(5),X(5),TS(20),T1(5),XO(5),ANG(5),H(5,5),EH(5,10
     1),AI(5,5)
      COMMON A,C,XO,CONT,KONT,N,KOUNT,PI,ITYPE,ANG,K1,ITIMES,H,EH,AI
      TA12=A(1,2)
      TA34=A(3,4)/TA12
      EEE=(A(1,1)-A(4,4))/TA12
      KAY=1
      E=ABSF(C(2))*SQRTF(X(1)*X(1)+X(2)*X(2))
      PHI=ANGLE(X(2)*C(2),X(1)*C(2))
      EE=ABSF(C(4))*SQRTF(X(3)*X(3)+X(4)*X(4))
      PHII=ANGLE(X(4)*C(4),X(3)*C(4))
      TIME=0.
      GO TO 28
41    IF(TEMP2)8,44,8
44    TIME=TIME+1.E-20
      GO TO 28
8     NS=0
      KAY=2
      D=SIGN(TEMP2)
      TE=TA12*T
      V=.5/(1.+TE)
      IF(TE-V)300,300,301
301   TB=9.*V/TE
      GO TO 6
300   TB=0.
      V=TE/10.
6     K=0
      TIME=0.0
      TEMP1=D
```

```
34      DUM=V+TB*TIME
        TIME=TIME+DUM
        IF(TIME-TE-DUM) 28,28,29
29      IF(NS)31,3,31
31      IF(TS(NS)-T)3,5,5
5       NS=NS-1
3       RETURN
28      TEMP2=EXPF(EEE*TIME)*E*SINF(TIME+PHI)+EE*SINF(TA34*TIME+PHII)
        GO TO (41,61,11),KAY
61      IF(TEMP1*TEMP2)30,30,34
30      K=K+1
        KAY=3
        NS=K
        IF(NS-50)1,1,101
101     PRINT 102
102     FORMAT(22H NS IS GREATER THAN 50)
        CALL EXIT
1       JT=0
        DINC=DUM/2.
        TIME=TIME-DINC
        IF(TEMP2)205,4,205
4       TIME=TIME+DINC
        GO TO 2
205     U=-SIGN(TEMP2)
10      JT=JT+1
        IF(JT-25)204,2,2
204     DINC=DINC/2.
        GO TO 28
11      IF(U*TEMP2)200,2,200
200     TIME=TIME+SIGN(U*TEMP2)*DINC
        GO TO 10
2       TS(K)=TIME/TA12
        TEMP1=-TEMP1
        KAY=2
        GO TO 34
        END
```

Subroutine CALXH(XG, NS, T, D, TS, XH, ENM, E)

CALXH calculates XH $= \mathbf{M}(\mathbf{v}_k, T_k)$, E $= \mathbf{x}_0 - \mathbf{M}(\mathbf{v}_k, T_k)$, and ENM $= \mathbf{x}_0 - \mathbf{M}(\mathbf{v}_k, T_k)$, given XG, NS, T, D, and TS.

```
        SUBROUTINE CALXH(XG,NS,T,D,TS,XH,ENM,E)
        DIMENSION XG(5),A(5,5),C(5),TS(20),XH(5),E(5),FUN(5,5),T1(5)
       1,AI(5,5),XO(5),Y(5),ANG(5),H(5,5),EH(5,10)
        COMMON A,C,XO,CONT,KONT,N,KOUNT,PI,ITYPE,ANG,K1,ITIMES,H,EH,AI
2       CALL FUND(A,T,FUN,N)
        CALL MLTV(FUN,N,N,XO,Y)
        CALL MLTV(FUN,N,N,C,T1)
        DD=(-1.)**NS
        DO 6 I=1,N
6       T1(I)=D*(T1(I)*DD-C(I))
        IF(NS)7,7,8
8       W=-2.*D
        DO 9 I=1,NS
        W=-W
        CALL FUND(A,TS(I),FUN,N)
        CALL MLTV(FUN,N,N,C,XH)
        DO 9 J=1,N
```

```
9       T1(J)=T1(J)+W*XH(J)
7       CALL MLTV(AI,N,N,T1,XH)
        DO 5 I=1,N
        XH(I)=XG(I)+XH(I)
5       F(I)=Y(I)-XH(I)
        CALL FUND(A,-T,FUN,N)
        CALL MLTV(FUN,N,N,F,T1)
        ENM=SQRTF(SCPROD(T1,T1,N))
        RETURN
        END
```

Subroutine CPRNT(NS, ENM, T, XG, TS, E, D, SCAL, XH)

CPRNT is used to print the appropriate data at the end of each iteration.

```
      SUBROUTINE CPRNT(NS,ENM,T,XG,TS,E,D,SCAL,XH)
      DIMENSIONA(5,5),C(5),XO(5),ANG(5),H(5,5),EH(5,10),XG(5),TS(20),E(5
     1),XH(5),AI(5,5),T1(5)
      COMMON A,C,XO,CONT,KONT,N,KOUNT,PI,ITYPE,ANG,K1,ITIMES,H,EH,AI
      IF(KOUNT)1,1,2
1     PUNCH 3
3     FORMAT(9H1A MATRIX)
      DO 4 I=1,N
4     PUNCH5,(A(I,J),J=1,N)
5     FORMAT(8F10.5)
      PUNCH 6
6     FORMAT(9H C VECTOR)
      PUNCH5,(C(I),I=1,N)
      PUNCH 1002,ITYPE,KONT,N,ANG(N),CONT
1002 FORMAT(7H/ITYPE=I4/17H ITERATION LIMIT=I4/7H ORDER=I4/8H RADIUS=E1
     12.4/15H NORM ACCURACY=E12.4)
      PUNCH 1005
1005  FORMAT(23H STATE TO BE CONTROLLED)
      PUNCH 5,(XO(I),I=1,N)
      PUNCH 8
8     FORMAT(3X,4HV(1),6X,4HV(2),6X,4HV(3),6X,4HV(4),4X,2HIT,1X,2HNS,5X,
     13HENM,8X,4HTIME,2X,3HSGN,1X,4HSTEP)
2     PUNCH9,(XG(I),I=1,N),KOUNT,NS,ENM,T,D,SCAL
9     FORMAT(4E10.4,I3,I3,1X,E10.4,1X,F8.4,1X,F4.1,1X,F6.4)
258   RETURN
      END
```

Subroutine CALG(NS, T, TS, XG, D, XH)

CALG calculates the elements of the matrix $\mathbf{G}(v_k, T_k)$.

```
      SUBROUTINE CALG(NS,T,TS,XG,D,XH)
      DIMENSIONTS(20),A(5,5),C(5),XG(5),TEMP(5,5),T1(5),T2(5),ANG(5),XH(
     15),XO(5),H(5,5),EH(5,10),AI(5,5)
      COMMON A,C,XO,CONT,KONT,N,KOUNT,PI,ITYPE,ANG,K1,ITIMES,H,EH,AI
      M=N
      DO 3 I=1,N
      DO 3 J=1,N
      K=J+N
      EH(I,K)=0.
3     TEMP(I,J)=0.
      CALL CALANG(XG,ANG,N)
```

```
801   IF(NS) 222,222,221
222   CALL CJACO(ANG,N,H)
      DO 1 I=1,N
      DO 1 J=1,K1
1     EH(I,J)=H(I,J)
      GO TO 223
221    DO 224 I=1,NS
      CALL CFUND(A,TS(I),H,N)
      CALL CMLTV(H,N,N,C,T1)
      CALL CMLTV(A,N,N,T1,T2)
      DTS=2./ABSF(CSPROD(T2,XG,N))
      CALL CSTV (T2,T1,DTS,N)
      DO 224 J=1,N
      DO 224 K=J,N
224   TEMP(J,K)=T1(J)*T2(K)+TEMP(J,K)
      DO 100 I=1,N
      TEMP(I,I)=TEMP(I,I)+1.
      DO 100 J=I,N
100   TEMP(J,I)=TEMP(I,J)
      CALL CJACO(ANG,N,H)
246   DO 101 I=1,M
      DO 101 J=1,K1
      EH(I,J)=0.
      DO 101 K=1,M
101   EH(I,J)=TEMP(I,K)*H(K,J)+EH(I,J)
C     DETERMINATION OF STATE VELOCITY
223   CALL CMLTV(A,N,N,XH,T1)
      DUM=D*(-1.0)**NS
      CALL CFUND(A,T,TEMP,N)
      CALL CMLTV(TEMP,N,N,C,T2)
      DO 235 I=1,M
      L=I+N
      EH(I,L)=1.
235   EH(I,M)=-T1(I)+DUM*T2(I)
      RETURN
      END
```

Subroutine CALNG(AL, T, XG, NS, D, TS, DW, ENM, E, XH)

CALNG calculates a new guess by determining the best step size γ_k (AL in CALNG). The method of Section 8.4 is used in a modified form. The value of AL is doubled each time that CALNG is called (statement 47) unless AL $\geqslant$ 1. The reason for this is that γ_k should approach 1 as k increases to infinity. Also, a lower bound of 5×10^{-4} was placed on γ_k (see the statement after 46). This bound was chosen arbitrarily, and it proved successful in several experiments where γ_k would have otherwise been smaller. The variable LO is used to keep track of the number of integrations, and these data are printed each time that CALNG is called.

```
      SUBROUTINE CALNG(AL,T,XG,NS,D,TS,DW,ENM,E,XH)
      DIMENSIONA(5,5),C(5),XG(5),ANG(5),TS(20),XO(5),DW(5),DXG(5),DTS(20
     1),DXH(5),DE(5),E(5),XH(5),GA(5,5),T1(5),H(5,5),EH(5,10),AI(5,5)
      COMMON A,C,XO,CONT,KONT,N,KOUNT,PI,ITYPE,ANG,K1,ITIMES,H,EH,AI
      LO=1
      CALLCMLTV(H,N,K1,DW,T1)
      IF(KOUNT)60,60,47
```

```
47    AL=AL*2.
      IF(AL-1.)44,44,60
60    AL=1.
44    DO 40 I=1,N
40    DXG(I)=XG(I)+AL*T1(I)
      DXGNM=ANG(N)/SQRTF(CSPROD(DXG,DXG,N))
      DT=T+AL*DW(N)
      DO 41 I=1,N
41    DXG(I)=DXG(I)*DXGNM
      CALLCTIMES(DXG,DT,DD,KNS,DTS)
      CALL CALXH(DXG,KNS,DT,DD,DTS,DXH,DENM,DE)
      GO TO (42,11,10),LO
42    IF(DENM-.95*ENM)10,10,9
11    IF(DENM-ENM)10,10,12
10    CALLCEQ(NS,KNS,D,DD,ENM,DENM,XG,DXG,XH,DXH,E,DE,TS,DTS,T,DT,N)
      PUNCH 13,LO
13    FORMAT(24H NUMBER OF INTEGRATIONS=I10)
      RETURN
9     LO=2
46    AL=.5*AL*AL*ENM/(DENM-(1.-AL)*ENM)
      IF(AL-.0005)45,44,44
45    AL=.0005
      GO TO 44
12    LO=3
      GO TO 46
      END
```

Subroutine CEQ(NS, KNS, D, DD, ENM, DENM, XG, DXG, XH, DXH, E, DEE, TS, DTS, T, DT, N)

CEQ is used to equate the variables in its argument by pairs (that is, NS = KNS, D = DD, etc.).

```
      SUBROUTINECEQ(NS,KNS,D,DD,ENM,DENM,XG,DXG,XH,DXH,E,DEE,TS,DTS,T,DT
     1,N)
      DIMENSIONXG(5),DXG(5),XH(5),DXH(5),E(5),DEE(5),TS(20),DTS(20)
      NS=KNS
      D=DD
      T=DT
      ENM=DENM
      DO 1 I=1,N
      XG(I)=DXG(I)
      XH(I)=DXH(I)
1     E(I)=DEE(I)
      IF(NS)2,2,3
2     RETURN
3     DO 4 I=1,NS
4     TS(I)=DTS(I)
      RETURN
      END
```

Subroutine CWORST(T, TJ, XG)

Given two times T and TJ, the matrix **A**, the vector **b**, and the radius r, CWORST finds a vector **v** in S, such that

$$\langle \mathbf{v}, \mathbf{A}^{m} e^{\mathbf{A} T_j} \mathbf{b} \rangle = 0, \quad m = 1, 2, \ldots, N-1,$$

where $T_j = TJ$. Also, $\mathbf{M}$ is evaluated on this pair [that is, $(\mathbf{v}, T)$] to obtain an initial state $\mathbf{x}_0$ whose corresponding optimal boundary condition is in Q_μ.

```
      SUBROUTINE CWORST(T,TJ,XG)
      DIMENSIONA(5,5),C(5),XG(5),XO(5),ANG(5),EH(5,10),T1(5),T2(5),T3(5)
     1,TS(20),H(5,5)
      COMMON A,C,XO,CONT,KONT,N,KOUNT,PI,ITYPE,ANG,K1,ITIMES,H,EH
      DO 8 I=1,N
      DO 8 J=1,N
      K=N+J
8     EH(I,K)=0.
      XG(N)=.5*ANG(N)
      CALL CFUND(A,TJ,H,N)
      CALL CMLTV(H,N,N,C,T1)
      T3(1)=-T1(N)*XG(N)
      DO 1 I=1,K1
      K=I+K1
      EH(I,K)=1.
1     EH(1,I)=T1(I)
      DO 2 I=2,K1
      CALL CMLTV(A,N,N,T1,T2)
      T3(I)=-T2(N)*XG(N)
      DO 2 J=1,N
      T1(J)=T2(J)
2     EH(I,J)=T2(J)
      CALL CLINEQ(EH,K1,2*K1,M)
      GO TO (4,3),M
4     PRINT 5
5     FORMAT(21H NO SOLUTION IN WORST)
      CALL EXIT
3     DO13 I=1,K1
      T1(I)=0.
      DO13 J=1,K1
      L=J+K1
13    T1(I)=EH(I,L)*T3(J)+T1(I)
      T1(N)=XG(N)
      DUM=ANG(N)/SQRTF(CSPROD(T1,T1,N))
      CALL CSTV(XG,T1,DUM,N)
      CALL CTIMES(XG,T,D,NS,TS)
      CALL CALXH(XG,NS,T,D,TS,T3,ENM,T1)
      CALL CFUND(A,-T,H,N)
      CALL CMLTV(H,N,N,T3,XO)
      RETURN
      END
```

Appendix G. The Program for Computing Fixed-Time Minimum-Fuel Controls

General Description

The list of FORTRAN statements used to obtain the experimental results of Section 8.5 is given in this appendix. The program is designed to calculate the fuel-optimal control that steers the state from a given initial condition $\mathbf{x}_0$ to a given hyperspherical target set at the origin of the state space in time T. The restrictions on the plant for using this program are the same as those described in Appendix F for the time-optimal program. The initial guess can be read in or calculated, and all of the subroutines, except TIMES, are written for any plant satisfying the restrictions just mentioned. The naming of the subroutines in this program corresponds, except for subroutines CHGAL and XSIMEQF, to the names in the minimum-time program (that is, subroutines performing the same functions have been given similar names). Also, wherever possible the same variable notation is used. Consequently, the description of each subroutine given here is brief. Subroutine XSIMEQF performs the same function as CLINEQ in Appendix F. Whereas the program of Appendix F is the 1620 version, the one listed here is written for the 7094.

Subroutine MAIN

This part of the program initiates the computation by reading in the pertinent data (that is, $\mathbf{A}$, T, $\mathbf{b}$, $\mathbf{x}_0$, etc.). The variables conform to the time-optimal MAIN with the exception of the following variables, which are new:

$$(\mathrm{AI(I, J), I = 1, N, J = 1, N}) = \mathbf{A}^{-1}, \tag{G.1}$$

$$(\mathrm{AM(I, J), I = 1, N, J = 1, N}) = e^{-\mathbf{A}T}, \tag{G.2}$$

$$(\mathrm{AT(I, J), I = 1, N, J = 1, N}) \quad = e^{\mathbf{A}T}, \tag{G. 3}$$

$$(\mathrm{AP(I, J), I = 1, N, J = 1, N}) \quad = e^{-(\mathbf{A}+\mathbf{A}')T}. \tag{G. 4}$$

These arrays are calculated and stored in COMMON before proceeding with the computation, since they remain unchanged during the iteration. The variable ITYPE controls the experiment. if ITYPE = 1, the program will read in the initial guess $(\mathbf{v}_1, \alpha_1)$. If ITYPE = 2, 3, or 4, then

$$\mathbf{v}_1 = \frac{re^{\mathbf{A}T}\mathbf{x}_0}{\|e^{\mathbf{A}T}\mathbf{x}_0\|} \tag{G. 5}$$

is calculated, and subroutine CHGAL ("change α") is called to select α(AL in the PROGRAM) so that for ITYPE = 4, $f(\mathbf{x}_0; \alpha_1) = 0$ as in the first procedure of Section 6. 6, for ITYPE = 3, $\langle \mathbf{x}_0 - \mathbf{M}(\mathbf{v}_1, \alpha_1), \mathbf{M}_\alpha(\mathbf{v}_1, \alpha_1)\rangle = 0$ as in the second procedure of Section 6. 6, and finally for ITYPE = 2, the cost $J(\mathbf{v}_1, \alpha_1)$ equals a read-in number DJ (the statement following 2). After the first guess is found, subroutine GRDINV is called to start the iterative procedure.

```
      DIMENSION A(10,10),AT(10,10),AI(10,10),AM(10,10),AP(10,10),EXA(10,
     150),C(10),XO(10),TT(10),H(10,10),G(10,10),ANG(10),XG(10),T1(10),TS
     2(50),DI(50),XH(10),E(10)
      COMMON A,AT,AI,AM,AP,EXA,C,XO,TT,H,G,K1,CD,KOUNT,ITYPE,N,T,ANG,CON
     1T,KONT
8     READ1,BLA,CONT,KONT,N,T,AL,ITYPE
      LIMIT=0
      K1=N-1
      ANG(N)=BLA
      DO 2 I=1,N
2     READ3,(A(I,J),J=1,N)
      READ3,(C(I),I=1,N),DJ
      DO 11 I=1,N
      DO 11 J=1,N
      AI(I,J)=A(I,J)
11    AT(I,J)=0.
      DO 12 I=1,N
12    AT(I,I)=1.
      DE=1.
      M=XSIMEQF(10,N,N,AI,AT,DE,T1)
      READ3,(T1(I),I=1,N)
      CALL FUND(A,-T,AM,N)
      CALL FUND(A,T,AT,N)
      CALL MLTV(AT,N,N,T1,XO)
      DO 5 I=1,N
      DO 5 J=I,N
      AP(I,J)=0.
      DO 5 K=1,N
      AP(I,J)=AM(I,K)*AM(J,K)+AP(I,J)
      AP(I,J)=AP(J,I)
      PRINT 4, BLA,CONT,KONT,N,T,AL,(T1(I),I=1,N)
      PRINT 17,ITYPE
```

```
17    FORMAT(7H ITYPE=I10)
      GO TO (14,13,13,13),ITYPE
14    READ 16,AL,(T1(I),I=1,N)
      GO TO 15
13    DUM=ANG(N)/SQRTF(SCPROD(XO,XO,N))
      IF(DUM-.999)7,6,6
6     PRINT 9
      GO TO 8
7     DO 10 I=1,N
10    T1(I)=XO(I)*DUM
      CALL TIMES(AL,T1,NS,TS,DI,IC)
      CALL CALXH(AL,T1,XH,E,TS,DI,ENM,IC,NS)
      DENM=ENM
      ALPHA=AL
      CALL CHGAL(AL,DJ,ALPHA,T1,XH,E,TS,DI,ENM,DENM,IC,NS,COST)
15    CALL GRDINV(AL,T1,XH,E,TS,DI,ENM,IC,NS)
      GO TO 8
1     FORMAT(2E10.1,2I10,2F10.4,I10)
16    FORMAT(8F10.5)
3     FORMAT(4F10.5)
4     FORMAT(8H RADIUS=E10.3/10H ACCURACY=E10.3/17H ITERATION LIMIT=I10/
     17H ORDER=I10/6H TIME=F10.4/4H AL=F10.4/7H STATE=4F12.5)
9     FORMAT(25 H OPTIMUM CONTROL IS U = 0)
      END
      FUNCTION COSTN(NS,IC,TS)
      DIMENSION TS(50)
      COSTN=T*FLOATF(XMODF(NS+IC+1,2))
      KK=NS+1
      W=(-1.)**(IC-1)
      DO 3 I=2,KK
      W=-W
3     COSTN=COSTN+W*TS(I)
      RETURN
      END
```

Subroutine CHGAL(AL, DJ, ALPHA, XG, XH, E, TS, DI, ENM, DENM, IC, NS, COST)

The function of this subroutine is to carry out a one-dimensional search in R_1^+ to find the zero of one of three possible functions of α (which are being determined by the variable ITYPE which is in COMMON). The three functions are as follows: $J(\mathbf{v}_1, \alpha) - DJ$, where DJ is data input to the subroutine; $\langle \mathbf{x}_0 - \mathbf{M}(\mathbf{v}_1, \alpha), \mathbf{M}_\alpha(\mathbf{v}_1, \alpha)\rangle$, which is evaluated by the function TESTN (see statements 2 and/or 8); or $f(\mathbf{x}_0, \alpha)$ (see statements 28 and/or 29). The function $J(\mathbf{v}_1, \alpha)$ is evaluated by the function COSTN (see statements 1 and/or 7). The search is carried out by incrementing α (AL) in step sizes ALPHA until the zero is bracketed. The bracketed interval (of length ALPHA) is then searched by continuously halving the interval and testing until the desired accuracy is reached.

```
      SUBROUTINE CHGAL(AL,DJ,ALPHA,XG,XH,E,TS,DI,ENM,DENM,IC,NS,COST)
      DIMENSION A(10,10),AT(10,10),AI(10,10),AM(10,10),AP(10,10),EXA(10,
     150),C(10),XO(10),TT(10),H(10,10),G(10,10),ANG(10),XG(10),XH(10),E(
     210),TS(50),DI(50)
      COMMON A,AT,AI,AM,AP,EXA,C,XO,TT,H,G,K1,CD,KOUNT,ITYPE,N,T,ANG,CON
     1T,KONT
      LIMIT=0
      KAY=1
      GO TO (1,1,2,28),ITYPE
1     CO=COSTN(NS,IC,TS)
      TEST=COST+DJ-CO
      PRINT 4,TEST,AL,CO
4     FORMAT(6H COST=E12.4,4H AL=E12.4,4H CO=E12.4)
      GO TO 3
2     TEST=TESTN(NS,DI,E,XG,AL,DENM)
      PRINT 5,TEST,AL
5     FORMAT(6H EMAL=E12.4,4H AL=E12.4)
      GO TO 3
28    TEST=SCPROD(XG,E,N)
      PRINT 30,TEST,AL
30    FORMAT(7H TESTN=E12.4,4H AL=E12.4)
3     IF(ABSF(TEST)-.0001)19,19,6
19    RETURN
6     IF(TEST)18,19,20
18    U=-1.
      AL=AL-ALPHA
      GO TO 21
20    U=1.
      AL =AL+ALPHA
21    DUM=DENM
      CALL TIMES(AL,XG,NS,TS,DI,IC)
      CALL CALXH(AL,XG,XH,E,TS,DI,DENM,IC,NS)
      IF(ABSF(DENM-DUM)-.001)19,19,27
27    GO TO (7,7,8,29),ITYPE
7     CO=COSTN(NS,IC,TS)
      TEST=COST+DJ-CO
      PRINT 4,TEST,AL,CO
      GO TO 9
8     TEST=TESTN(NS,DI,E,XG,AL,DENM)
      PRINT 5,TEST,AL
      GO TO 9
29    TEST=SCPROD(XG,E,N)
      PRINT 30,TEST,AL
9     PRINT 10,DENM,ENM,ALPHA
10    FORMAT(6H DENM=E12.4,5H ENM=E12.4,7H ALPHA=E12.4)
      GO TO (25,23),KAY
25    IF(U*TEST)23,19,6
23    ALPHA=ALPHA/2.
      IF(ALPHA-.001)19,19,26
26    KAY=2
      LIMIT=LIMIT+1
      IF(LIMIT-10)6,19,19
      END
```

Subroutine GRDINV(AL, XG, XH, E, TS, DI, ENM, IC, NS)

GRDINV is used to control the iteration. This is done in the same manner as in MAIN in the program of Appendix F.

```
      SUBROUTINE GRDINV(AL,XG,XH,E,TS,DI,ENM,IC,NS)
      DIMENSION A(10,10),AT(10,10),AI(10,10),AM(10,10),AP(10,10),EXA(10,
     150),C(10),XO(10),TT(10),H(10,10),G(10,10),ANG(10),XG(10),DW(10),VE
     2L(10),T1(10),XH(10),E(10),TS(50),PA(10,10),FUN(10,10),DI(50)
      COMMON A,AT,AI,AM,AP,EXA,C,XO,TT,H,G,K1,CD,KOUNT,ITYPE,N,T,ANG,CON
     1T,KONT
      KOUNT=0
      GO TO (4,30,30,30),ITYPE
4     CALL TIMES(AL,XG,NS,TS,DI,IC)
5     CALL CALXH(AL,XG,XH,E,TS,DI,ENM,IC,NS)
30    CALL PRNTPC(AL,NS,ENM,XG,TS,E,DI,IC,XH)
      IF(ENM-CONT)8,8,9
9     IF(KOUNT-KONT)17,8,8
8     RETURN
17    CALL CALG(AL,DI,XG,NS)
      GO TO 22
21    CALL MLTV(PA,N,N,VEL,DW)
      PRINT 24
24    FORMAT(25H GOING INTO GRADIENT MODE)
      GO TO 26
22    DO 70 I=1,N
      VEL(I)=E(I)
      DO 70 J=1,N
70    PA(I,J)=G(J,I)
      DE=1.
      M=XSIMEQF(10,N,1,G,E,DE,T1)
      GO TO (285,21,21),M
285   DO 29 I=1,N
29    DW(I)=G(I,1)
26    PRINT 702,(DW(I),I=1,N)
702   FORMAT(6H DW = 10E12.5)
25     CALL CALNG(AL,XG,NS,DI,TS,DW,ENM,E,XH,IC)
      KOUNT=KOUNT+1
      GO TO 30
      END
```

Subroutine TIMES(AL, XG, NS, TS, DI, IC)

Subroutine TIMES is used to calculate $\eta(\mathbf{v}, \alpha)$, $\mu(\mathbf{v}, \alpha)$, $w_j(\mathbf{v}, \alpha)$, $j = 1, 2, \ldots, \mu(\mathbf{v}, \alpha)$, $j = 1, 2, \ldots, \mu(\mathbf{v}, \alpha)$. The switch times are found by searching the interval (0, T) for the zeros of $|\langle \alpha\mathbf{v}, e^{A\tau}\mathbf{b}\rangle| - 1$. This subroutine is very similar to subroutine CTIMES in Appendix F.

```
      SUBROUTINE TIMES(AL,XG,NS,TS,DI,IC)
      DIMENSION A(10,10),AT(10,10),AI(10,10),AM(10,10),AP(10,10),EXA(10,
     150),C(10),XO(10),TT(10),H(10,10),G(10,10),ANG(10),XG(10),DI(50),TS
     2(50)
      COMMON A,AT,AI,AM,AP,EXA,C,XO,TT,H,G,K1,CD,KOUNT,ITYPE,N,T,ANG,CON
     1T,KONT
      LIMIT=0
      TS(1)=0.
      NS=0
```

```
      PI=3.1415927
      E1=A(1,1)/A(1,2)
      E2=A(4,4)/A(1,2)
      TA34=A(3,4)/A(1,2)
44    TEMP1=AL*(XG(2)*C(2)+XG(4)*C(4))
      TEMP2=TEMP1
       DUM=TEMP1
      IF(ABSF(TEMP1)-1.)35,36,37
35    IC=1
      GO TO 6
37    IC=2
      GO TO 6
36    DIM=AL*(C(2)*(XG(1)*A(1,2)+XG(2)*A(1,1))+C(4)*(XG(3)*A(3,4)+XG(4)*
     1A(4,4)))
      IF(DIM*TEMP1)35,38,37
38    PRINT 39
      CALL EXIT
6     TE=T*A(1,2)
      V=.3/(1.+TE)
      WW=V
      TEMP1=SIGNF(1.,ABSF(TEMP1)-1.)
      IF(TEMP1)76,77,76
77    TEMP1=SIGNF(1.,DIM*DUM)
76    K=0
      TI=0.
      E=ABSF(C(2))*SQRTF(XG(1)**2+XG(2)**2)*AL
      EE=ABSF(C(4)*SQRTF(XG(3)**2+XG(4)**2)*AL)
      PHI=ANGLE(XG(2)*C(2),XG(1)*C(2))
      PHII=ANGLE(XG(4)*C(4),XG(3)*C(4))
      TA=TE+V*(TE+1.)
34    DUM=V*(1.+TI)
      TI=TI+DUM
      W=SIGNF(1.,TEMP2)
      IF(TI-TA)28,28,29
29    IF(TS(NS+1)-T)12,11,11
11    NS=NS-1
12    TS(NS+2)=T
      DI(NS+2)=DI(NS+1)
      DI(1)=DI(2)
      IF(NS)40,40,41
41    RETURN
40    PRINT 33,IC
33    FORMAT(34H NO OF SWITCHINGS ZERO CHGE AL IC=I10)
      GO TO (43,42),IC
43    AL=1.1*AL
      GO TO 44
42    AL=.9*AL
      GO TO 44
28    TEMP2=EXPF(E1*TI)*E*SINF(TI+PHI)+EE*SINF(TA34*TI+PHII)*EXPF(E2*TI)
100   TANP=ABSF(TEMP2)-1.
1     IF(TEMP1*TANP)30,30,70
70    IF(TANP)34,30,75
75    IF(W*TEMP2)71,34,34
71    V=V/10.
      LIMIT=LIMIT+1
      IF(LIMIT-3)72,73,73
73    PRINT 74,DUM,TI,TANP,TEMP2,W,V,Y
```

```
74    FORMAT(28H STEP SIZE PROBLEMS IN TIMES7E12.4)
      CALL EXIT
72     TI=TI-DUM
      TEMP2=-TEMP2
      DUM=V*(1.+TI)
      GO TO 34
30    K=K+1
      NS=K
      V=WW
      LIMIT=0
      IF(NS-48)1,1,101
101   PRINT 102
      CALL EXIT
1     JT=0
      DINC=DUM/2.
      TI=TI-DINC
      IF(TANP)205,4,206
4     TI=TI+DINC
      GO TO 2
206   U=-1.
      GO TO 10
205   U=1.
10    JT=JT+1
      IF(JT-20)204,2,2
204   DINC=DINC/2.
      TEMP2=EXPF(E1*TI)*E*SINF(TI+PHI)+EE*SINF(TA34*TI+PHII)*EXPF(E2*TI)
      TANP=ABSF(TEMP2)-1.
      IF(U*TANP)200,2,201
200   TI=TI-DINC
      GO TO 10
201   TI=TI+DINC
      GO TO 10
2     L=K+1
      TS(L)=TI/A(1,2)
      DI(L)=SIGNF(1.,TEMP2)
      TEMP1=-TEMP1
      GO TO 34
102   FORMAT(16H NS IS TOO LARGE)
39    FORMAT(17H TROUBLE IN TIMES)
      END
```

Subroutine CALXH(AL, XG, Y, E, TS, DI, ENM, IC, NS)

Subroutine CALXH evaluates $e^{\mathbf{A}T}\mathbf{M}(\mathbf{v}, \alpha)$ [(Y(I), I = 1, N) in the subroutine], $e^{\mathbf{A}T}(\mathbf{x}_0 - \mathbf{M}(\mathbf{v}, \alpha))$ [(E(I), I = 1, N] in the subroutine, $\|\mathbf{x}_0 - \mathbf{M}(\mathbf{v}, \alpha)\|$ (ENM in the subroutine), and the vectors $e^{\mathbf{A}\tau_j}\mathbf{b}$, u = 0, 1, 2, ..., $\mu + 2$ [(EXA(I, J), I = 1, N, J = 1, NS + 2) in the program]. The information stored in EXA is used in subroutine CALG and function TESTN.

```
      SUBROUTINE CALXH(AL,XG,Y,E,TS,DI,ENM,IC,NS)
      DIMENSION A(10,10),AT(10,10),AI(10,10),AM(10,10),AP(10,10),EXA(10,
     150),C(10),XO(10),TT(10),H(10,10),G(10,10),ANG(10),T1(10),Y(10),E(1
     20),TS(50),DI(50),DN(10,10),XG(10)
      COMMON A,AT,AI,AM,AP,EXA,C,XO,TT,H,G,K1,CD,KOUNT,ITYPE,N,T,ANG,CON
     1T,KONT
      II=2-IC
      IJ=1-XMODF(NS+IC,2)
      IE=NS+IJ+IC-1
      DO 1 I=1,N
```

```
1     E(I)=0.
      W=1.
      DO 2 I=1,IE
      W=-W
      K=I+II
      CALL FUND(A,TS(K),DN,N)
      CALL MLTV(DN,N,N,C,T1)
      DO 5 L=1,N
5     EXA(L,K)=T1(L)
      DO 2 J=1,N
2     E(J)=W*T1(J)*DI(I+II)+E(J)
      CALL MLTV(AI,N,N,E,T1)
      DO 3 I=1,N
      Y(I)=T1(I)+XG(I)
3     E(I)=XO(I)-Y(I)
      CALL MLTV(AM,N,N,E,T1)
      ENM=SQRTF(SCPROD(T1,T1,N))
      RETURN
      END
```

Subroutine CALG(AL, DI, XG, NS)

As in the minimum-time program, subroutine CALG evaluates the matrix $\mathbf{G}(\mathbf{v}, \alpha)$ ((G(I, J), I = 1, N, J = 1, N) in the program. The vector $(1/J_\alpha(\mathbf{v}, \alpha))\, \mathbf{J}_\mathbf{v}(\mathbf{v}, \alpha)$ (see Equation 1.16) is also stored in the array (TT(I), I = 1, N) in COMMON, as is

$$CD = \alpha J_\alpha(\mathbf{v}, \alpha).$$

These data are used in subroutine CALNG.

```
      SUBROUTINE CALG(AL,DI,XG,NS)
      DIMENSION A(10,10),AT(10,10),AI(10,10),AM(10,10),AP(10,10),EXA(10,
     150),C(10),XO(10),TT(10),H(10,10),G(10,10),ANG(10),XG(10),DI(50),FU
     2N(10,10),QL(50,10),STD(50),DUM(50),AQL(10,50)
      COMMON A,AT,AI,AM,AP,EXA,C,XO,TT,H,G,K1,CD,KOUNT,ITYPE,N,T,ANG,CON
     1T,KONT
      IF(NS-1)10,10,11
11    INS=NS+1
      DO 7 I=1,N
      DO 7 J=2,INS
      AQL(I,J)=0.
      DO 7 K=1,N
7     AQL(I,J)=A(I,K)*EXA(K,J)+AQL(I,J)
      CD=0.
      DO 1 I=1,NS
      STD(I)=0.
      DO 2 K=1,N
2     STD(I)=XG(K)*AQL(K,I+1)+STD(I)
      STD(I)=1./ABS(AL*STD(I))
      CD=STD(I)+CD
      DO 1 K=1,N
      DUM(I)=STD(I)*DI(I+1)
1     QL(I,K)=AL*EXA(K,I+1)*STD(I)
```

```
100   CDP=AL/CD
      DO 3 I=1,N
      TT(I)=0.
      DO 3 J=1,NS
3     TT(I)=EXA(I,J+1)*DUM(J)+TT(I)
      DO 4 I=1,N
      DO 4 J=I,N
4     H(I,J)=-TT(I)*TT(J)*CDP
      DO 8 I=1,N
      DO 8 J=I,N
      FUN(I,J)=0.
      DO 8 K=1,NS
8     FUN(I,J)=EXA(I,K+1)*QL(K,J)+FUN(I,J)
102   CDP=AL*AL/CD
      DO 5 I=1,N
      G(I,N)=TT(I)/CD
      TT(I)=CDP*TT(I)
      FUN(I,I)=1.+FUN(I,I)
      DO 5 J=I,N
      FUN(I,J)=FUN(I,J)+H(I,J)
5     FUN(J,I)=FUN(I,J)
10    CALL CALANG(XG,ANG,N)
      CALL JACO(ANG,N,H)
      CDP=AL*AL*DI(2)
      DO 15 I=1,N
15    TT(I)=CDP*EXA(I,2)
      IF(NS-1)12,12,13
13    DO 9 I=1,N
      DO 9 J=1,K1
      G(I,J)=0.
      DO 9 K=1,N
9     G(I,J)=FUN(I,K)*H(K,J)+G(I,J)
      RETURN
12    DO 14 I=1,N
      G(I,N)=EXA(I,2)
      DO 14 J=1,K1
14    G(I,J)=H(I,J)
101   RETURN
      END
```

Subroutine CALNG(AL, XG, NS, DI, TS, DW, ENM, E, XH, IC)

This subroutine is almost identical to its counterpart in Appendix F.

```
      SUBROUTINE CALNG(AL,XG,NS,DI,TS,DW,ENM,E,XH,IC)
      DIMENSION A(10,10),AT(10,10),AI(10,10),AM(10,10),AP(10,10),EXA(10,
     150),C(10),XO(10),TT(10),H(10,10),G(10,10),ANG(10),XG(10),DI(50),TS
     2(50),DW(10),E(10),XH(10),DXG(10),DTS(50),DXH(10),DE(10),DDI(50),T1
     3(10)
      COMMON A,AT,AI,AM,AP,EXA,C,XO,TT,H,G,K1,CD,KOUNT,ITYPE,N,T,ANG,CON
     1T,KONT
      LO=1
      COST=COSTN(NS,IC,TS)
      CALL MLTV(H,N,K1,DW,T1)
      DW(N)=DW(N)*AL/CD-SCPROD(TT,T1,N)
      PRINT 17,DW(N)
```

```
17    FORMAT(7H DW(N)=E12.4)
      IF(ABSF(DW(N))-.5*AL)60,60,61
61    DAL=.5*AL/ABSF(DW(N))
      GO TO 62
60    DAL=1.
62    DO 40 I=1,N
40    DXG(I)=XG(I)+DAL*T1(I)
      DXGNM=ANG(N)/SQRTF(SCPROD(DXG,DXG,N))
      DL=AL+DAL*DW(N)
      ALPHA=ABSF(DAL*DW(N))
      DO 41 I=1,N
41    DXG(I)=DXGNM*DXG(I)
      CALL TIMES(DL,DXG,KNS,DTS,DDI,KIC)
      CALL CALXH(DL,DXG,DXH,DE,DTS,DDI,DENM,KIC,KNS)
      GO TO (42,11,10),LO
42    IF(DENM-.5*ENM)10,10,9
10    CALLEQ(NS,KNS,DI,DDI,ENM,DENM,XG,DXG,XH,DXH,E,DE,TS,DTS,AL,DL,IC,
     1KIC)
      PRINT 13,LO,DAL
13    FORMAT(4H LO=I10,5H DAL=F10.5)
      RETURN
9     LO=2
      DAL=DAL*.5*ENM/DENM
      GO TO 62
11    IF(DENM-ENM)10,12,12
12    LO = 3
      AL=.5*AL*AL*ENM/(DENM-(1.-AL)*ENM)
      GO TO 44
      END
```

The two functions TESTN and COSTN and the subroutine EQ, next listed, complete the listing of the program for calculating fixed-time fuel-optimal controls. These subprograms are considered to be self-explanatory. COSTN evaluates the fuel used by a given dead-zone control, and TESTN calculates $\langle \mathbf{e}_k, \mathbf{M}_\alpha(\mathbf{v}_k, \alpha_k)\rangle$. Subroutine EQ performs the same function here as CEQ did in Appendix F.

```
      FUNCTION TESTN(NS,DI,E,XG,AL,DENM)
      DIMENSION A(10,10),AT(10,10),AI(10,10),AM(10,10),AP(10,10),EXA(10,
     150),C(10),XO(10),TT(10),H(10,10),G(10,10),ANG(10),T3(10),DI(50),T2
     2(10),E(10),XG(10),T4(10)
      COMMON A,AT,AI,AM,AP,EXA,C,XO,TT,H,G,K1,CD,KOUNT,ITYPE,N,T,ANG,CON
     1T,KONT
      AL2=AL*AL*DENM
      TESTN=0.
      DO 41 J=1,NS
      DO 40 I=1,N
40    T4(I)=EXA(I,J+1)
      CALL MLTV(AP,N,N,T4,T2)
      CALL MLTV(A,N,N,T4,T3)
41    TESTN=TESTN+DI(J+1)*SCPROD(E,T2,N)/(AL2*ABSF(SCPROD(XG,T3,N)))
      RETURN
      END
```

```
      FUNCTION COSTN(NS,IC,TS)
      DIMENSION TS(50)
      COSTN=T*FLOATF(XMODF(NS+IC+1,2))
      KK=NS+1
      W=(-1.)**(IC-1)
      DO 3 I=2,KK
      W=-W
3     COSTN=COSTN+W*TS(I)
      RETURN
      END


      SUBROUTINEEQ(NS,KNS,DI,DDI,ENM,DENM,XG,DXG,XH,DXH,E,DE,TS,DTS,AL,D
     1AL,IC,KIC)
      DIMENSION A(10,10),AT(10,10),AI(10,10),AM(10,10),AP(10,10),EXA(10,
     150),C(10),XO(10),TT(10),H(10,10),G(10,10),ANG(10),XG(10),DXG(10),X
     2H(10),DXH(10),E(10),DE(10),TS(50),DTS(50),DI(50),DDI(50)
      COMMON A,AT,AI,AM,AP,EXA,C,XO,TT,H,G,K1,CD,KOUNT,ITYPE,N,T,ANG,CON
     1T,KONT
10    NS=KNS
      IC=KIC
      K=NS+1
      ENM=DENM
      DO 1000 I=1,N
      XG(I)=DXG(I)
      XH(I)=DXH(I)
1000  E(I)=DE(I)
      DO 1001 I=1,K
      DI(I)=DDI(I)
1001  TS(I)=DTS(I)
      AL=DAL
      J=K+1
      TS(J)=DTS(J)
      RETURN
      END


      SUBROUTINE PRNTPC(AL,NS,ENM,XG,TS,E,DI,IC,XH)
      DIMENSION A(10,10),AT(10,10),AI(10,10),AM(10,10),AP(10,10),EXA(10,
     150),C(10),XO(10),TT(10),H(10,10),G(10,10),ANG(10),XG(10),TS(50),E(
     210),DI(50),T1(10),XH(10)
      COMMON A,AT,AI,AM,AP,EXA,C,XO,TT,H,G,K1,CD,KOUNT,ITYPE,N,T,ANG,CON
     1T,KONT
      COST=COSTN(NS,IC,TS)
      K=NS+1
      IF(KOUNT)1,1,2
1     PRINT 3
3     FORMAT(9H1A MATRIX)
      DO 4 I=1,N
4     PRINT 5,(A(I,J),J=1,N)
5     FORMAT(10F10.5)
      PRINT 6
6     FORMAT(9H C VECTOR)
      PRINT 5,(C(I),I=1,N)
      PRINT 8
8     FORMAT(1H0,24X,9HITERATION,3X,2HNS,5X,3HENM,10X,2HAL,4X,6HU-TYPE,4
     1X,4HCOST)
2     PRINT 9,KOUNT,NS,ENM,AL,IC,COST
```

```
9     FORMAT(1H ,24X,I9,3X,I2,E12.4,E12.4,1X,I5,5X,F7.4)
      CALL MLTV(AM,N,N,XH,T1)
      PRINT 251


251   FORMAT(1H+,82X,7HCOSTATE)
      PRINT 252,(XG(I),I=1,N)
252    FORMAT(1H ,80X,4E12.5)
      PRINT 254
254    FORMAT(1H ,82X,4HXHAT)
      PRINT 252,(XH(I),I=1,N)
      PRINT 252,(T1(I),I=1,N)
      PRINT 255
255    FORMAT(1H ,82X,5HERROR)
      PRINT 252,(E(I),I=1,N)
      PRINT 256
256    FORMAT(1H ,82X,6HANGLES)
      PRINT 252,(ANG(I),I=1,K1)
      PRINT 257
257   FORMAT(1H ,82X,12HSWITCH TIMES)
      PRINT 252,(TS(I),I=2,K)
      PRINT 260
260   FORMAT(1H ,82X,22HCONTROLS AT SWITCHINGS)
      PRINT 252,(DI(I),I=2,K)
      RETURN
      END
```

Appendix H. A Program for Solving the Terminal-Cost Problem when the Effort is not Constrained

The program listed in this appendix is a realization of the algorithm of Section 2. 3 and was written for use on the Disc Version of the IBM 1620 computer. However, it was found that it was not necessary to use the disc when the variables were DIMENSIONED as listed here. The experiments described in Section 8. 2 were carried out using this program.

The variables in the program may be associated with those of Section 2. 3 as follows:

N . . . . n

M . . . . m

KONT . . . . k

T . . . . T

A(I, J) . . . . $\mathbf{A}$

AI(I, J) . . . . $\mathbf{A}^{-1}$

Q(I, J) . . . . $\mathbf{Q}$

B(I, J) . . . . $\mathbf{B}$

XO(I) . . . . x_0

Z(1) . . . . z_k

D(I, J) . . . . $e^{\mathbf{A}T}$

XII(I) . . . . $e^{\mathbf{A}T}x_0$

ATB(I, J) . . . . $e^{\%T}\mathbf{B}$

DLA . . . . λ_k

Y(I) . . . . y_k

XH(I) . . . . $\bar{x}_k(T)$

W(I) . . . . w_k

XG(I) . . . . $p_k(T)$

The so-called MAIN of the program is used to read the data and to carry out the basic steps of the algorithm. The input data are n, m, KOUNT, T, CONT, ITYPE, $\mathbf{A}$, $\mathbf{Q}$, $\mathbf{B}$, x_0, and z. KOUNT is used as an upper limit on the number of iterations, CONT is a lower limit on λ, and ITYPE is used to control whether the initial guess is read in or calculated. (Controlling the initial-guess procedure in this manner is convenient for restarting an experiment already partially completed.)

The variable TA in the program is used to represent the maximum of the magnitudes of the elements of the **A** matrix, and the array SA(I, J) is simply **A** normalized by this number. The reason for defining SA is to have a basis for choosing the step size in CTIME when searching for the switch times.

The matrix AI, that is, the inverse of **A** (see CLINEQ), and the matrices e^{AT} and $e^{AT}\mathbf{B}$ are determined in the MAIN for use elsewhere in the program (for example, in CALXHT). Subroutine CALXHT is used to find $\bar{\mathbf{x}}_k(T)$.

It should be noted that this program is written to include the case where m is greater than one (that is, the multi-input case).

The reader is reminded that any subroutine called by this program, which is not listed in this appendix, will be found listed in one of the others.

```
      COMMONA(5,5),B(5,5),Q(5,5),D(5,5),XO(5),X1(5),XH(5),Z(5),Y(5),XG(5
     1),NS(5),TS(5,30),DC(5),X11(5),AI(5,5),ATB(5,5),EH(5,10),N,T,CONT,
     2KOUNT,M,KONT,DLA,ENM,DENM,TA,SA(5,5)
      DIMENSION W(5),QW(5)
10    READ6,N,M,KOUNT,T,CONT,ITYPE
      DO 7 I=1,N
7     READ8,(A(I,J),J=1,N)
      TA=ABS(A(1,2))
      DO 70 I=1,N
      DO 70 J=1,N
      IF(ABS(A(I,J))-TA)70,70,71
71    TA=ABS(A(I,J))
70    CONTINUE
      DO 72 I=1,N
      DO 72 J=1,N
72    SA(I,J)=A(I,J)/TA
      DO 31 I=1,N
      DO 31 J=1,N
      EH(I,J)=A(I,J)
31    EH(I,J+N)=0.
      DO 41 I=1,N
41    EH(I,I+N)=1.
      CALL CLINEQ(N,2*N)
      DO 42 I=1,N
      DO 42 J=1,N
42    AI(I,J)=EH(I,J+N)
      DO 14 I=1,N
14    READ 8,(Q(I,J),J=1,N)
      DO 15 I=1,M
15    READ 8,(B(J,I),J=1,N)
      READ 8,(XO(I),I=1,N)
      READ 8, (Z(I),I=1,N)
      KONT=0
      CALL CFUND(A,T,D,N)
      CALL CMLTV(D,N,N,XO,X11)
      DO 50 I=1,N
      DO 50 J=1,M
      ATB(I,J)=0.
      DO 50 K=1,N
```

```
50    ATB(I,J)=D(I,K)*B(K,J)+ATB(I,J)
      GO TO (60,61),ITYPE
61    READ8,(X1(I),I=1,N)
      GO TO 4
60    DO 18 I=1,N
18    X1(I)=X11(I)
      GO TO 4
5     P=1.-DLA
        DO 3 I=1,N
3     X1(I)=P*X1(I)+DLA*XH(I)
      KONT=KONT+1
      IF(KONT-KOUNT)4,4,9
4     DO 1 I=1,N
1     Y(I)=X1(I)-Z(I)
      ENM=SQRTF(CSPROD(Y,Y,N))
      CALL CMLTV(Q,N,N,Y,XG)
      CALL CALXHT
      DO 2 I=1,N
2     W(I)=XH(I)-X1(I)
      IF(CSPROD(W,W,N)-1.E-3)9,9,22
22    CALL CMLTV(Q,N,N,W,QW)
      DLA=(CSPROD(Y,QW,N)/CSPROD(W,QW,N))*(-1.)
      IF(DLA-1.)16,11,17
17    DLA=1.
      GO TO 11
16    IF(DLA)19,19,11
11    CALL CPRNT3
      GO TO 5
19    PUNCH 21
21    FORMAT(14HDLA IS NOT +VE)
9     CALL CPRNT3
      GO TO 10
6     FORMAT(3I10,F10.5,E10.1,I10)
8     FORMAT(8F10.5)
      END
```

```
      SUBROUTINE CALXHT
      COMMONA(5,5),B(5,5),Q(5,5),D(5,5),XO(5),X1(5),XH(5),Z(5),Y(5),XG(5
     1),NS(5),TS(5,30),DC(5),X11(5),AI(5,5),ATB(5,5),EH(5,10),N,T,CONT,
     2KOUNT,M,KONT,DLA,ENM,DENM,TA,SA(5,5)
      DIMENSION T1(5),FUN(5,5)
      DO 1 I=1,N
1     T1(I)=0.
      DO 7 JK=1,M
      CALL CTIMT(JK)
      DD=(-1)**NS(JK)
      DO 6 I=1,N
6     T1(I)=DC(JK)*(ATB(I,JK)*DD-B(I,JK))+T1(I)
      IF(NS(JK))7,7,8
8     W=-2.*DC(JK)
      DO 9 I=1,NS(JK)
      W=-W
      CALL CFUND(A,TS(JK,I),FUN,N)
      DO 2 J=1,N
      XH(J)=0.
      DO 2 K=1,N
```

```
2     XH(J)=FUN(J,K)*B(K,JK)+XH(J)
      DO 9 J=1,N
9     T1(J)=T1(J)+W*XH(J)
7     CONTINUE
      CALL CMLTV(AI,N,N,T1,XH)
      DO 3 I=1,N
      XH(I)=X11(I)-XH(I)
3     T1(I)=Z(I)-XH(I)
      DENM=SQRT(CSPROD(T1,T1,N))
      RETURN
      END


      SUBROUTINE CTIMT(JK)
      COMMONA(5,5),B(5,5),Q(5,5),D(5,5),XO(5),X1(5),XH(5),Z(5),Y(5),XG(5
     1),NS(5),TS(5,30),DC(5),X11(5),AI(5,5),ATB(5,5),EH(5,10),N,T,CONT,
     2KOUNT,M,KONT,DLA,ENM,DENM,TA,SA(5,5)
      DIMENSION SD(5,5),TI(5)
      KAY=1
      TI=0.
      NS(JK)=0
      TE=TA*T
      V=.5/(1.+TE)
      IF(TE-V)300,300,301
300   V=TE/10.
      TB=0.
      GO TO 28
301   TB=9.*V/TE
      GO TO 28
8     IF(TEMP2)6,4,6
4     TI=TI+.00001
      GO TO 28
6     TI=0.
      KAY=2
      TEMP1=SIGN(TEMP2)
      DC(JK)=TEMP1
      DUM=V
34    DUM=V+TB*TI
      TI=TI+DUM
      IF(TI-TE-DUM)28,28,29
      IF(NS)31,3,31
31    IF(TS(JK,NS(JK))-T)3,5,5
5     NS(JK)=NS(JK)-1
3     RETURN
28    CALL CFUND(SA,TI,SD,N)
      TEMP2=0.
      DO 70 I=1,N
      DO 70 J=1,N
70    TEMP2=TEMP2+SD(I,J)*B(J,JK)*XG(I)
      GO TO (8,61,11),KAY

61    IF(TEMP2*TEMP1)30,30,34
30    KAY=3
      NS(JK)=NS(JK)+1
      IF(NS(JK)-30)1,1,101
101   PRINT 102
102   FORMAT(22H NS IS GREATER THAN 30)
      CALL EXIT
```

```
 1     JT=0
       DINC=DUM/2.
       TI=TI-DINC
       U=-SIGN(TEMP2)
 10    JT=JT+1
       IF(JT-15)204,2,2
 204   DINC=DINC/2.
       GO TO 28
 11    IF(U*TEMP2)200,2,200
 200   TI=TI+SIGN(U*TEMP2)*DINC
       GO TO 10
 2     TS(JK,NS(JK))=TI/TA
       TEMP1=-TEMP1
       KAY=2
       GO TO 34
       END


       SUBROUTINE CPRNT3
       COMMONA(5,5),B(5,5),Q(5,5),D(5,5),XO(5),X1(5),XH(5),Z(5),Y(5),XG(5
      1),NS(5),TS(5,30),DC(5),X11(5),AI(5,5),ATB(5,5),EH(5,10),N,T,CONT,
      2KOUNT,M,KONT,DLA,ENM,DENM,TA,SA(5,5)
       IF(KONT)1,1,2
1      PUNCH21
       DO 3 I=1,N
3      PUNCH4,(A(I,J),J=1,N)
       PUNCH 5
       DO 6 I=1,M
6      PUNCH4,(B(J,I),J=1,N)
       PUNCH 7
       DO 8 I=1,N
8      PUNCH4,(Q(I,J),J=1,N)
       PUNCH 9
       PUNCH 4,(XO(I),I=1,N)
       PUNCH 10
       PUNCH 4,( Z(I),I=1,N)
       PUNCH 11,N,M,KOUNT,T,CONT
2      PUNCH 12,KONT,DLA,ENM,DENM
       PUNCH 13,(X1(I),I=1,N)
       PUNCH 14,(Y(I),I=1,N)
       PUNCH 15,(XG(I),I=1,N)
       PUNCH 16,(XH(I),I=1,N)
       PUNCH 17,(NS(I),I=1,M)
       DO 18 I=1,M
       L=NS(I)
       IF(L-1)18,23,23
23     PUNCH 19,(TS(I,J),J=1,L)
18     CONTINUE
21     FORMAT(9H A MATRIX)
4      FORMAT(8F10.5)
 5      FORMAT(9H B MATRIX)
 7      FORMAT(9H Q MATRIX)
9      FORMAT(14H INITIAL STATE)
10     FORMAT(13H TARGET STATE)
11     FORMAT(16H ORDER OF PLANT=I10/18H NUMBER OF INPUTS=I10/17H ITERATI
      1ON LIMIT=I10/18H CONTROL INTERVAL=F10.5/18H ACCURACY CONTROL=E9.2)
12     FORMAT(10HITERATION=I3,3X,4HDLA=F7.4,3X,4HENM=F8.4,3X,5HDENM=F8.4)
13     FORMAT(3HX1=6E12.4)
```

```
14    FORMAT(3H Y=6E12.4)
15    FORMAT(3HXG=6E12.4)
16    FORMAT(3HXH=8E10.3)
17    FORMAT(4HNS =5I5)
19    FORMAT(3HTS=6E12.4)
      RETURN
      END
```

```
      COMMON N,T,CONT,KOUNT,M,A(5,5),B(5,2),Q(5,5),XO(5),Z(5),D(5,5),X1(
     15),XH(5),XG(5),KONT,XJ,YJ,DLA,AL,NS(2),TS(2,40),DI(2,40),IC(2),ENM
     2,DENM,X11(5),AI(5,5),Y(5),EH(5,10)
      DIMENSION W(5),QW(5)
      AL=10
10    READ6,N,M,KOUNT,T,XJ,CONT,ITYPE,IDATA
      DO 7 I=1,N
7     READ8,(A(I,J),J=1,N)
      DO 31 I=1,N
      DO 31 J=1,N
      L=J+N
      EH(I,J)=A(I,J)
31    EH(I,L)=0.
      DO 41 I=1,N
      K=I+N
41    EH(I,K)=1.
      CALL CLINEQ(N,2*N)
      DO 42 I=1,N
      DO 42 J=1,N
      K=J+N
42    AI(I,J)=EH(I,K)
      DO 14 I=1,N
14    READ 8,(Q(I,J),J=1,N)
      DO 15 I=1,M
15    READ 8,(B(J,I),J=1,N)
      READ 8,(XO(I),I=1,N)
      READ 8, (Z(I),I=1,N)
      KONT=0
      CALL CFUND(A,T,D,N)
      CALL CMLTV(D,N,N,XO,X11)
      GO TO (45,46),IDATA
46    READ 8,(X1(I),I=1,N)
      GO TO 4
45    DO 18 I=1,N
18    X1(I)=X11(I)
      GO TO 4
5     P=1.-DLA

      DO 3 I=1,N
3     X1(I)=P*X1(I)+DLA*XH(I)
      KONT=KONT+1
      IF(KONT-KOUNT)4,4,9
4     DO 1 I=1,N
1     Y(I)=X1(I)-Z(I)
      ENM=SQRTF(CSPROD(Y,Y,N))
      CALL CMLTV(Q,N,N,Y,XG)
      CALL CALX1H
      DO 2 I=1,N
```

```
2     W(I)=XH(I)-X1(I)
      IF(CSPROD(W,W,N)-1.E-5)9,9,22
22    CALL CMLTV(Q,N,N,W,QW)
      DLA=(CSPROD(Y,QW,N)/CSPROD(W,QW,N))*(-1.)
      IF(DLA-1.)16,11,17
17    DLA=1.
      GO TO 11
16    IF(DLA)19,19,11
11    CALL CPRNT2
      GO TO 5
19    PUNCH 21
21    FORMAT(14HDLA IS NOT +VE)
9     CALL CPRNT2
      KONT=0
      GO TO (10,44),ITYPE
44    READ 8,TT,TXJ
      T=T+TT
      XJ=XJ+TXJ
      PUNCH 47,T,XJ
47    FORMAT(6HNEW T=F10.4,7HNEW XJ=F10.4)
      GO TO 4
6     FORMAT(3I10,2F10.5,E10.1,2I10)
8     FORMAT(8F10.5)
      END
```

```
      SUBROUTINE CTIME(JK)
      COMMON N,T,CONT,KOUNT,M,A(5,5),B(5,2),Q(5,5),XO(5),Z(5),D(5,5),X1(
     15),XH(5),XG(5),KONT,XJ,YJ,DLA,AL,NS(2),TS(2,40),DI(2,40),IC(2),ENM
     2,DENM,X11(5),AI(5,5),Y(5),EH(5,10)
      LIMIT=0
      KAY=1
      TI=0.
      TS(JK,1)=0.
      NS(JK)=0
      E1=A(1,1)/A(1,2)
      E2=A(4,4)/A(1,2)
      TA34=A(3,4)/A(1,2)
      E3=ABSF(B(2,JK))*SQRTF(XG(1)*XG(1)+XG(2)*XG(2))*AL
      EE=ABSF(B(4,JK)*SQRTF(XG(3)*XG(3)+XG(4)*XG(4))*AL)
      PHI=ANGLE(XG(2)*B(2,JK),XG(1)*B(2,JK))
      PHII=ANGLE(XG(4)*B(4,JK),XG(3)*B(4,JK))
      GO TO 28
44    IF(ABSF(TEMP2)-1.)35,36,37
35    IC(JK)=1
      DI(JK,1)=0.
      GO TO 6
37    IC(JK)=2
      DI(JK,1)=SIGN(TEMP2)
      GO TO 6
36    TI=TI+.00001
      GO TO 28
6     TE=T*A(1,2)
      V=.2/(1.+TE)
      KAY=2
      WW=V
```

```
      TEMP1=SIGN(ABSF(TEMP2)-1.)
      TI=0.
      TA=TE+V
34    DUM=V*(1.+TI)
      TI=TI+DUM
      WWW=SIGN(TEMP2)
      IF(TI-TA)28,28,29
29    IF(TS(JK,L)-T)12,11,11
11    NS(JK)=NS(JK)-1
12    LS=NS(JK)+2
      TS(JK,LS)=T
      DI(JK,LS)=.5*(1.-(-1.)**(NS(JK)+IC(JK)+1))*DI(JK,LS-1)
41    RETURN
28    TEMP2=EXP(E1*TI)*E3*SIN(TI+PHI)+EE*SIN(TA34*TI+PHII)*EXP(E2*TI)
100   TANP=SIGN(ABSF(TEMP2)-1.)
      GO TO (44,61,77),KAY
61    IF(TEMP1*TANP)30,30,70
70    IF(TANP)34,30,75
75    IF(WWW*TEMP2)71,34,34
71    V=V/10.
      LIMIT=LIMIT+1
      IF(LIMIT-4)72,73,73
73    TYPE 74
74    FORMAT(28H STEP SIZE PROBLEMS IN CTIME)
      CALL EXIT
72    TI=TI-DUM
      TEMP2=-TEMP2
      DUM=V*(1.+TI)
      GO TO 34
30    NS(JK)=NS(JK)+1
      KAY=3
      V=WW
      LIMIT=0
      IF(NS(JK)-38)1,1,101
101   TYPE 102
      CALL EXIT
1     JT=0
      DINC=DUM/2.
      TI =TI-DINC
      U=-TANP
10    JT=JT+1
      IF(JT-15)204,2,2
204   DINC=DINC/2.
      GO TO 28
77    IF (TANP*U) 200,2,200
200   TI=TI+SIGN(U*TANP)*DINC
      GO TO 10
2     L=NS(JK)+1
      TS(JK,L)=TI/A(1,2)
      DI(JK,L)=SIGN(TEMP2)
      TEMP1=-TEMP1
      KAY=2
      GO TO 34
102   FORMAT(16H NS IS TOO LARGE)
      END
```

```
      SUBROUTINE CALX1H
      COMMON N,T,CONT,KOUNT,M,A(5,5),B(5,2),Q(5,5),XO(5),Z(5),D(5,5),X1(
     15),XH(5),XG(5),KONT,XJ,YJ,DLA,AL,NS(2),TS(2,40),DI(2,40),IC(2),ENM
     2,DENM,X11(5),AI(5,5),Y(5),EH(5,10)
      JT=0
      KAY=1
      DUM=10
3     YJ=0.
      DO 4 I=1,M
      CALL CTIME(I)
4     YJ=YJ+COSTN(I)
      TEMP=SIGN(XJ-YJ)
11    GO TO (5,9),KAY
5     DUM=2.*DUM
6     AL=AL+DUM*TEMP
      JT=JT+1
      IF(AL) 13,13,7
13    DUM=DUM*.5
      AL=.5*(AL-2.*TEMP*DUM)
7     YJ=0.
      DO 8 I=1,M
      CALL CTIME(I)
8     YJ=YJ+COSTN(I)
      JT=JT+1
      TEMP1=SIGN(XJ-YJ)
      IF(TEMP*TEMP1)9,10,11
9     DUM=DUM/2.
      IF(ABSF(XJ-YJ)-CONT)10,10,12
12    AL=AL+TEMP1*DUM
      KAY=2
      GO TO 7
10    CALL CALH
      TYPE 14,JT
      RETURN
14    FORMAT(33HNO OF CALCULATIONS OF SWITCHINGS=I10)
      END
```

```
      SUBROUTINE CALH
      COMMON N,T,CONT,KOUNT,M,A(5,5),B(5,2),Q(5,5),XO(5),Z(5),D(5,5),X1(
     15),XH(5),XG(5),KONT,XJ,YJ,DLA,AL,NS(2),TS(2,40),DI(2,40),IC(2),ENM
     2,DENM,X11(5),AI(5,5),Y(5),EH(5,10)
      DIMENSION DN(5,5),E(5),T2(5)
      DO 1 I=1,N
1     E(I)=0.
      DO 4 JK=1,M
      II=2-IC(JK)
      IE=NS(JK)+IC(JK)-(1-(-1)**(NS(JK)+IC(JK)))/2
      WW=1.
      DO 4 I=1,IE
      WW=-WW
      CALL CFUND(A,TS(JK,I+II),DN,N)
      DO 8 L=1,N
      T2(L)=0.
      DO 8 J=1,N
```

```
8      T2(L)=DN(L,J)*B(J,JK)+T2(L)
       DO 4 J=1,N
4      E(J)=T2(J)*WW*DI(JK,I+II)+E(J)
       CALL CMLTV(AI,N,N,E,T2)
       DO 2 I=1,N
       XH(I)=X11(I)-T2(I)
2      E(I)=XH(I)-Z(I)
       DENM=SQRTF(CSPROD(E,E,N))
       RETURN
       END
```

```
       SUBROUTINE CPRNT2
       COMMON N,T,CONT,KOUNT,M,A(5,5),B(5,2),Q(5,5),XO(5),Z(5),D(5,5),X1(
      15),XH(5),XG(5),KONT,XJ,YJ,DLA,AL,NS(2),TS(2,40),DI(2,40),IC(2),ENM
      2,DENM,X11(5),AI(5,5),Y(5),EH(5,10)
       IF(KONT)1,1,2
 1     PUNCH21
       DO 3 I=1,N
 3     PUNCH4,(A(I,J),J=1,N)
       PUNCH 5
       DO 6 I-1,M
 6     PUNCH4,(B(J,I),J=1,N)
       PUNCH 7
       DO 8 I=1,N
 8     PUNCH4,(Q(I,J),J=1,N)
       PUNCH 9
       PUNCH 4,(XO(I),I=1,N)
       PUNCH 10
       PUNCH 4,( Z(I),I=1,N)
       PUNCH 11,N,M,KOUNT,T,XJ,CONT
 2     PUNCH 12,KONT,DLA,YJ,ENM,DENM,AL
       PUNCH 13,(X1(I),I=1,N)
       PUNCH 14,(Y(I),I=1,N)
       PUNCH 15,(XG(I),I=1,N)
       PUNCH 16,(XH(I),I=1,N)
       PUNCH 17,(NS(I),I=1,M)
       DO 18 I=1,M
       L=NS(I)+2
       PUNCH 19,(TS(I,J),J=1,L)
 18    PUNCH 20,(DI(I,J),J=1,L)
 21    FORMAT(9H A MATRIX)
 4     FORMAT(8F10.5)
 5     FORMAT(9H B MATRIX)
 7     FORMAT(9H Q MATRIX)
 9     FORMAT(14H INITIAL STATE)
 10    FORMAT(13H TARGET STATE)
 11    FORMAT(16H ORDER OF PLANT=I10/18H NUMBER OF INPUTS=I10/17H ITERATI
      1ON LIMIT=I10/18H CONTROL INTERVAL=F10.5/19H EFFORT CONSTRAINT=F10.
      25/18H ACCURACY CONTROL=E9.2)
 12    FORMAT(11HITERATION =I3,3X,5HDLA =F7.4,3X,6HCOST =F7.4,3X,5HENM =F
      18.4/6HDENM =F8.4,3X,4HAL =E11.4)
 13    FORMAT(3HX1=6E12.4)
 14    FORMAT(3H Y=6E12.4)
```

```
15    FORMAT(3HXG=6E12.4)
16    FORMAT(3HXH=8E10.3)
17    FORMAT(4HNS =5I5)
19    FORMAT(3HTS=6E12.4)
20    FORMAT(4HDI =18F4.1)
22    RETURN
      END
```

Appendix I. A Program for Solving the Terminal-Cost Problem when there is Fuel Constraint on the Control.

The program listed in this appendix is a realization of the algorithm of Section 3. 4 for (1) $g(\mathbf{u}) = \sum_{i=1}^{m} |u_i|$ and (2) $\Omega = (\mathbf{u}; |u_i| \leq 1, \ i = 1, 2, \ldots)$. The experiments reported in Section 8. 3 were carried out using this program.

The procedure of Section 3. 4 is very similar to that of Section 2. 3 and, in fact, differs (that is, for this particular case) only in that one must adjust α (AL in the program) until the fuel used is equal to the constraint. The reader should note that it is not necessary, in this case, to carry out Step 2 of the iterative procedure.

The variable names in the program listed here have the same association as those of Appendix H where applicable. Otherwise, they can be interpreted as in Appendix C.

The MAIN program performs the same function as in Appendix H: subroutine CALX1H carries out Step 3; CALH integrates the plant equation in reverse time given the extremal controls; CTIME calculates the switch times and is similar to the subroutine by the same name in Appendix G, except that it is written for the case where there is more than one input; and CPRNT2 is used to print the data.

```
      COMMON N,T,CONT,KOUNT,M,A(5,5),B(5,2),Q(5,5),XO(5),Z(5),D(5,5),X1(
     15),XH(5),XG(5),KONT,XJ,YJ,DLA,AL,NS(2),TS(2,40),DI(2,40),IC(2),ENM
     2,DENM,X11(5),AI(5,5),Y(5),EH(5,10)
      DIMENSION W(5),QW(5)
      AL=10
 10   READ6,N,M,KOUNT,T,XJ,CONT,ITYPE,IDATA
      DO 7 I=1,N
```

```
7     READ8,(A(I,J),J=1,N)
      DO 31 I=1,N
      DO 31 J=1,N
      L=J+N
      EH(I,J)=A(I,J)
31    EH(I,L)=0.
      DO 41 I=1,N
      K=I+N
41    EH(I,K)=1.
      CALL CLINEQ(N,2*N)
      DO 42 I=1,N
      DO 42 J=1,N
      K=J+N
42    AI(I,J)=EH(I,K)
      DO 14 I=1,N
14    READ 8,(Q(I,J),J=1,N)
      DO 15 I=1,M
15    READ 8,(B(J,I),J=1,N)
      READ 8,(XO(I),I=1,N)
      READ 8, (Z(I),I=1,N)
      KONT=0
      CALL CFUND(A,T,D,N)
      CALL CMLTV(D,N,N,XO,X11)
      GO TO (45,46),IDATA
46    READ 8,(X1(I),I=1,N)
      GO TO 4
45    DO 18 I=1,N
18    X1(I)=X11(I)
      GO TO 4
5     P=1.-DLA
      DO 3 I=1,N
3     X1(I)=P*X1(I)+DLA*XH(I)
      KONT=KONT+1
      IF(KONT-KOUNT)4,4,9
4     DO 1 I=1,N
1     Y(I)=X1(I)-Z(I)
      ENM=SQRTF(CSPROD(Y,Y,N))
      CALL CMLTV(Q,N,N,Y,XG)
      CALL CALX1H
      DO 2 I=1,N
2     W(I)=XH(I)-X1(I)
      IF(CSPROD(W,W,N)-1.E-5)9,9,22
22    CALL CMLTV(Q,N,N,W,QW)
      DLA=(CSPROD(Y,QW,N)/CSPROD(W,QW,N))*(-1.)
      IF(DLA-1.)16,11,17
17    DLA=1.
      GO TO 11
16    IF(DLA)19,19,11
11    CALL CPRNT2
      GO TO 5
19    PUNCH 21
21    FORMAT(14HDLA IS NOT +VE)
9     CALL CPRNT2
      KONT=0
      GO TO (10,44),ITYPE
44    READ 8,TT,TXJ
      T=T+TT
      XJ=XJ+TXJ
      PUNCH 47,T,XJ
```

```
47    FORMAT(6HNEW T=F10.4,7HNEW XJ=F10.4)
      GO TO 4
6     FORMAT(3I10,2F10.5,E10.1,2I10)
8     FORMAT(8F10.5)
      END


      SUBROUTINE CTIME(JK)
      COMMON N,T,CONT,KOUNT,M,A(5,5),B(5,2),Q(5,5),XO(5),Z(5),D(5,5),X1(
     15),XH(5),XG(5),KONT,XJ,YJ,DLA,AL,NS(2),TS(2,40),DI(2,40),IC(2),ENM
     2,DENM,X11(5),AI(5,5),Y(5),EH(5,10)
      LIMIT=0
      KAY=1
      TI=0.
      TS(JK,1)=0.
      NS(JK)=0
      E1=A(1,1)/A(1,2)
      E2=A(4,4)/A(1,2)
      TA34=A(3,4)/A(1,2)
      E3=ABSF(B(2,JK))*SQRTF(XG(1)*XG(1)+XG(2)*XG(2))*AL
      EE=ABSF(B(4,JK)*SQRTF(XG(3)*XG(3)+XG(4)*XG(4))*AL)
      PHI=ANGLE(XG(2)*B(2,JK),XG(1)*B(2,JK))
      PHII=ANGLE(XG(4)*B(4,JK),XG(3)*B(4,JK))
      GO TO 28
44    IF(ABSF(TEMP2)-1.)35,36,37
35    IC(JK)=1
      DI(JK,1)=0.
      GO TO 6
37    IC(JK)=2
      DI(JK,1)=SIGN(TEMP2)
      GO TO 6
36    TI=TI+.00001
      GO TO 28
6     TE=T*A(1,2)
      V=.2/(1.+TE)
      KAY=2
      WW=V
      TEMP1=SIGN(ABSF(TEMP2)-1.)
      TI=0.
      TA=TE+V
34    DUM=V*(1.+TI)
      TI=TI+DUM
      WWW=SIGN(TEMP2)
      IF(TI-TA)28,28,29
29    IF(TS(JK,L)-T)12,11,11
11    NS(JK)=NS(JK)-1
12    LS=NS(JK)+2
      TS(JK,LS)=T
      DI(JK,LS)=.5*(1.-(-1.)**(NS(JK)+IC(JK)+1))*DI(JK,LS-1)
41    RETURN
28    TEMP2=EXP(E1*TI)*E3*SIN(TI+PHI)+EE*SIN(TA34*TI+PHII)*EXP(E2*TI)
100   TANP=SIGN(ABSF(TEMP2)-1.)
      GO TO (44,61,77),KAY
61    IF(TEMP1*TANP)30,30,70
70    IF(TANP)34,30,75
75    IF(WWW*TEMP2)71,34,34
```

```
71    V=V/10.
      LIMIT=LIMIT+1
      IF(LIMIT-4)72,73,73
73    TYPE 74
74    FORMAT(28H STEP SIZE PROBLEMS IN CTIME)
      CALL EXIT
72    TI=TI-DUM
      TEMP2=-TEMP2
      DUM=V*(1.+TI)
      GO TO 34
30    NS(JK)=NS(JK)+1
      KAY=3
      V=WW
      LIMIT=0
      IF(NS(JK)-38)1,1,101
101   TYPE 102
      CALL EXIT
1     JT=0
      DINC=DUM/2.
      TI =TI-DINC
      U=-TANP
10    JT=JT+1
      IF(JT-15)204,2,2
204   DINC=DINC/2.
      GO TO 28
77    IF (TANP*U) 200,2,200
200   TI=TI+SIGN(U*TANP)*DINC
      GO TO 10
2     L=NS(JK)+1
      TS(JK,L)=TI/A(1,2)
      DI(JK,L)=SIGN(TEMP2)
      TEMP1=-TEMP1
      KAY=2
      GO TO 34
102   FORMAT(16H NS IS TOO LARGE)
      END
```

```
      SUBROUTINE CPRNT2
      COMMON N,T,CONT,KOUNT,M,A(5,5),B(5,2),Q(5,5),XO(5),Z(5),D(5,5),X1(
     15),XH(5),XG(5),KONT,XJ,YJ,DLA,AL,NS(2),TS(2,40),DI(2,40),IC(2),ENM
     2,DENM,X11(5),AI(5,5),Y(5),EH(5,10)
      IF(KONT)1,1,2
1     PUNCH21
      DO 3 I=1,N
3     PUNCH4,(A(I,J),J=1,N)
      PUNCH 5
      DO 6 I=1,M
6     PUNCH4,(B(J,I),J=1,N)
      PUNCH 7
      DO 8 I=1,N
8     PUNCH4,(Q(I,J),J=1,N)
      PUNCH 9
      PUNCH 4,(XO(I),I=1,N)
      PUNCH 10
      PUNCH 4,( Z(I),I=1,N)
      PUNCH 11,N,M,KOUNT,T,XJ,CONT
```

```
2     PUNCH 12,KONT,DLA,YJ,ENM,DENM,AL
      PUNCH 13,(X1(I),I=1,N)
      PUNCH 14,(Y(I),I=1,N)
      PUNCH 15,(XG(I),I=1,N)
      PUNCH 16,(XH(I),I=1,N)
      PUNCH 17,(NS(I),I=1,M)
      DO 18 I=1,M
      L=NS(I)+2
      PUNCH 19,(TS(I,J),J=1,L)
18    PUNCH 20,(DI(I,J),J=1,L)
21    FORMAT(9H A MATRIX)
4     FORMAT(8F10.5)
5     FORMAT(9H B MATRIX)
7     FORMAT(9H Q MATRIX)
9     FORMAT(14H INITIAL STATE)
10    FORMAT(13H TARGET STATE)
11    FORMAT(16H ORDER OF PLANT=I10/18H NUMBER OF INPUTS=I10/17H ITERATI
     1ON LIMIT=I10/18H CONTROL INTERVAL=F10.5/19H EFFORT CONSTRAINT=F10.
     25/18H ACCURACY CONTROL=E9.2)
12    FORMAT(11HITERATION =I3,3X,5HDLA =F7.4,3X,6HCOST =F7.4,3X,5HENM =F
     18.4/6HDENM =F8.4,3X,4HAL =E11.4)
13    FORMAT(3HX1=6E12.4)
14    FORMAT(3H Y=6E12.4)
15    FORMAT(3HXG=6E12.4)
16    FORMAT(3HXH=8E10.3)
17    FORMAT(4HNS =5I5)
19    FORMAT(3HTS=6E12.4)
20    FORMAT(4HDI =18F4.1)
22    RETURN
      END


      SUBROUTINE CALX1H
      COMMON N,T,CONT,KOUNT,M,A(5,5),B(5,2),Q(5,5),XO(5),Z(5),D(5,5),X1(
     15),XH(5),XG(5),KONT,XJ,YJ,DLA,AL,NS(2),TS(2,40),DI(2,40),IC(2),ENM
     2,DENM,X11(5),AI(5,5),Y(5),EH(5,10)
      JT=0
      KAY=1
      DUM=10
3     YJ=0.
      DO 4 I=1,M
      CALL CTIME(I)
4     YJ=YJ+COSTN(I)
      TEMP=SIGN(XJ-YJ)
11    GO TO (5,9),KAY
5     DUM=2.*DUM
6     AL=AL+DUM*TEMP
      JT=JT+1
      IF(AL) 13,13,7
13    DUM=DUM*.5
      AL=.5*(AL-2.*TEMP*DUM)
7     YJ=0.
      DO 8 I=1,M
      CALL CTIME(I)
8     YJ=YJ+COSTN(I)
      JT=JT+1
      TEMP1=SIGN(XJ-YJ)
      IF(TEMP*TEMP1)9,10,11
```

```
9     DUM=DUM/2.
      IF(ABSF(XJ-YJ)-CONT)10,10,12
12    AL=AL+TEMP1*DUM
      KAY=2
      GO TO 7
10    CALL CALH
      TYPE 14,JT
      RETURN
14    FORMAT(33HNO OF CALCULATIONS OF SWITCHINGS=I10)
      END
```

```
      SUBROUTINE CALH
      COMMON N,T,CONT,KOUNT,M,A(5,5),B(5,2),Q(5,5),XO(5),Z(5),D(5,5),X1(
     15),XH(5),XG(5),KONT,XJ,YJ,DLA,AL,NS(2),TS(2,40),DI(2,40),IC(2),ENM
     2,DENM,X11(5),AI(5,5),Y(5),EH(5,10)
      DIMENSION DN(5,5),E(5),T2(5)
      DO 1 I=1,N
1     E(I)=0.
      DO 4 JK=1,M
      II=2-IC(JK)
      IE=NS(JK)+IC(JK)-(1-(-1)**(NS(JK)+IC(JK)))/2
      WW=1.
      DO 4 I=1,IE
      WW=-WW
      CALL CFUND(A,TS(JK,I+II),DN,N)
      DO 8 L=1,N
      T2(L)=0.
      DO 8 J=1,N
8     T2(L)=DN(L,J)*B(J,JK)+T2(L)
      DO 4 J=1,N
4     E(J)=T2(J)*WW*DI(JK,I+II)+E(J)
      CALL CMLTV(AI,N,N,E,T2)
      DO 2 I=1,N
      XH(I)=X11(I)-T2(I)
2     E(I)=XH(I)-Z(I)
      DENM=SQRTF(CSPROD(E,E,N))
      RETURN
      END
```

REFERENCES

1. Apostol, T. M., Principles of Mathematical Analysis, Addison-Wesley Publishing Co., Inc., Reading, Mass., 1960.

2. Athans, M., "On the Uniqueness of the Extremal Controls for a Class of Minimum Fuel Problems," IEEE Trans. Auto. Control, **AC-9**, 4 (1964).

3. Athans, M., and P. L. Falb, Optimal Control. An Introduction to the Theory and Its Applications, McGraw-Hill Book Co., New York, 1966.

4. Athans, M., and P. L. Falb, "Time-Optimal Control of a Class of Nonlinear Systems," IEEE Trans. Auto. Control, **AC-8**, 379 (1963)

5. Athans, M., P. L. Falb, and R. T. Lacoss, "Time Fuel, and Energy-Optimal Control of Nonlinear Norm Invariant Systems," IEEE Trans. Auto. Control, **AC-8**, 196–202 (1963).

6. Athans, M., P. L. Falb, and R. T. Lacoss, "On Optimal Control of Self Adjoint Systems" (preprints for the 1963 JACC).

7. Bartle, R. G., "Newton's Method in Banach Spaces," Proc. Amer. Math. Soc., **6**, 827-831 (1955).

8. Bass, R. W., "Optimal Feedback Control System Design by the Adjoint System," Technical Report 60-22A, Aeronica, Baltimore, Md., June 1960.

9. Bellman, R. E., and S. E. Dreyfus, Applied Dynamic Programming, Princeton University Press, Princeton, N.J., 1962.

10. Bellman, R., I. Glicksberg, and O. Gross, "On the Bang-Bang Control Problem," Quart. J. Appl. Mech. Math., **14**, 11–18 (1956).

11. Craig, A. J., and I. Flügge-Lotz, "Investigation of Optimal Control with Minimum-Fuel Consumption Criterion for a Fourth-Order Plant with Two Control Inputs; Synthesis of an Efficient Sub-Optimal Control," 1964 Joint Automatic Control Conference, Stanford, Calif., 207–222.

12. Dreyfus, S.E., "Dynamic Programming and the Calculus of Variations," J. Math. Anal. Appl., **1**, 228–239 (1960).

13. Eaton, J. H., "An Iterative Solution to Time-Optimal Control," J. Math. Anal. Appl., **5**, 329-344 (1962).

14. Fadden, E. J., and E. G. Gilbert, "Computational Aspects of the Time-Optimal Control Problem," in Computing Methods in Optimization Problems (A.V. Balakrishnan, L. W. Neustadt, eds.), Academic Press, New York, 1964.

15. Fancher, P. S., "Iterative Computation Procedures for an Optimum Control Problem," IEEE Trans. Auto. Control, **AC-10**, 346-348 (1965).

16. Gantmacher, F. R., The Theory of Matrices, Vols. I and II, Chelsea Publishing Co., New York, 1960.

17. Gilbert, Elmer, G., "An Iterative Procedure for Computing the Minimum of a Quadratic Form on a Convex Set," J. SIAM Control, 4, No. 1, 1966.

18. Harvey, C. A., and E. B. Lee, "On the Uniqueness of Time-Optimal Control for Linear Processes," J. Math. Anal. Appl., 5, 258-268 (1962).

19. Ho, Y. C., "A Successive Approximation Technique for Optimal Control Systems Subject to Input Saturation," Trans. ASME, J. Basic Eng., Series D, **84**, 33–40 (1962).

20. Kalman, R. E., and J. E. Bertram, "Control System Analysis and Design via the Second Method of Liapunov," Trans. ASME, J. Basic Eng., Series D, **82**, 371–393 (1960).

21. Kleinman, D. L. "Fuel Optimum Control of Second- and Third-Order Linear Systems with Different Time Constraints, S. M. Thesis, M.I.T., Cambridge, Mass., June, 1963.

22. Knudsen, H. K., "An Iterative Procedure for Computing Time-Optimal Control," IEEE Trans. Auto. Control, AC-9, 23-30 (1964).

23. Kolmogorov, A., and S. Fomin, Elements of the Theory of Functions and Functional Analysis, Vol. 1, Metric and Normed Spaces, Graylock Press, Rochester, N.Y., 1957.

24. LaSalle, J. P., "The Time-Optimal Control Problem," in Contributions to Differential Equations, Vol. V, Princeton University Press, Princeton, N.J., 1960.

25. Lee, E. B., "A Sufficient Condition in the Theory of Optimal Control," J. SIAM, Ser. A Control, 1, 241-245 (1963).

26. Lee, E. B., and L. Marcus, "Synthesis of Optimal Control for Nonlinear Processes with One Degree of Freedom," Inst. Math., Acad. of Sciences of the Ukrainian SSR, Kiev, 1961.

27. Lee, E. B., and L. Marcus, "Optimal Control for Nonlinear Processes," Archives for Rational Mechanics and Analysis, **8**, 36–58 (1961).

28. LeMay, J. L., "Recoverable and Reachable Zones for Control Systems with Linear Plants and Bounded Controller Outputs," IEEE Trans. Auto. Control, **AC-9**, 346–354 (1964).

29. Neustadt, L. W., "Synthesizing Time-Optimal Control Systems," J. Math. Anal. Appl., 1, 454–493 (1960).

30. Neustadt, L. W., "Time-Optimal Control Systems with Position and Integral Limits," J. Math. Anal. Appl., 3, 402-427 (1961).

31. Neustadt, L. W., "Minimum Effort Control Systems," J. SIAM, Ser. A Control, 1, 16–31 (1962).

32. Neustadt, L. W., "The Existence of Optimum Controls in the Absence of Convexity Conditions," J. Math. Anal. Appl., **7**, 110–117 (1963).

33. Paiewonsky, B., "Time-Optimal Control of Linear systems with Bounded Control," in International Symposium on Nonlinear Differential Equations and Nonlinear Mechanics, Academic Press, New York, 1963.

34. Paiewonsky, B., and L. W. Neustadt, "On Synthesizing Optimal Controls," Proc. of the Second IFAC Congress, Basle, 421/1–421/8 (1963).

35. Paiewonsky, B., *et al.*, "A Study of Synthesis Techniques for Optimal Plants," Technical Documentary Report, **ASD-TDR-63-239** (1964).

36. Plant, J. B., "An Iterative Procedure for the Computation of Fixed-Time Fuel-Optimal Controls," IEEE Trans. Auto. Control, **AC-11**, 652–660 (1966).

37. Plant, J. B., and M. Athans, "An Iterative Procedure for the Computation of Time-Optimal Controls," Proceedings of the Third IFAC Congress, London, Paper 13D (1965).

38. Pontryagin, L. S., V. Boltyanskii, R. Gamkrelidze, and E. Mishenko, The Mathematical Theory of Optimal Processes, Interscience Publishers, New York, 1962.

39. Roxin, E., "The Existence of Optimal Controls," Michigan Math. J., **9**, 109–119 (1962).

40. Todd, J., Survey of Numerical Analysis, McGraw-Hill Book Co., New York, 1962.

41. Zadeh, L., and C. Desoer, Linear System Theory, The State Space Approach, McGraw-Hill Book Co., New York, 1963.

Index